SpringerBriefs in Molecular Science

Green Chemistry for Sustainability

Series editor

Sanjay K. Sharma, Jaipur, India

More information about this series at http://www.springer.com/series/10045

Huamin Zhang · Xianfeng Li
Hongzhang Zhang

Li–S and Li–O$_2$ Batteries with High Specific Energy

Research and Development

 Springer

Huamin Zhang
Dalian National Laboratory for Clean
　　Energy, Dalian Institute of Chemical
　　Physics
Chinese Academy of Science
Dalian
China

Xianfeng Li
Dalian National Laboratory for Clean
　　Energy, Dalian Institute of Chemical
　　Physics
Chinese Academy of Science
Dalian
China

Hongzhang Zhang
Dalian National Laboratory for Clean
　　Energy, Dalian Institute of Chemical
　　Physics
Chinese Academy of Science
Dalian
China

ISSN 2191-5407　　　　　　　ISSN 2191-5415　(electronic)
SpringerBriefs in Molecular Science
ISSN 2212-9898
SpringerBriefs in Green Chemistry for Sustainability
ISBN 978-981-10-0744-6　　　ISBN 978-981-10-0746-0　(eBook)
DOI 10.1007/978-981-10-0746-0

Library of Congress Control Number: 2016954627

Printed on acid-free paper

This Springer imprint is published by Springer Nature
The registered company is Springer Nature Singapore Pte Ltd.
The registered company address is: 152 Beach Road, #22-06/08 Gateway East, Singapore 189721, Singapore

Preface

Li–S and Li–O$_2$ batteries own super high specific energy, 2600 Wh/kg for Li–S battery and 13000 Wh/kg for Li–O$_2$, respectively, which is several times higher than that of the commercial Li-ion batteries. These batteries could enable the electric vehicles or unmanned aerial crafts to have longer driving range, as well as make it possible to manufacture lighter and smaller portable electric devices. Currently, the research of Li–S and Li–O$_2$ batteries is among the hottest topics in energy storage and they are deemed as the next-generation energy storage devices beyond the Li-ion batteries. Many countries, including China, Korea, Japan, British and the USA, all provide large amounts of research funding to promote the development of Li–S and Li–O$_2$ batteries.

During the past decades, the Li–S and Li–O$_2$ batteries have achieved great progress in both fundamental research and application demonstration, and the possible solutions to many technical problems have also been proposed. This book offers a comprehensive overview of the fundamentals, recent developments, challenges and prospects of Li–S and Li–O$_2$ batteries, including the fundamental research and potential applications. The book illustrates the cell assembly, diagnostic test and electrolyte decomposition mechanism of Li–S and Li–O$_2$ batteries, and focuses on the development of key materials of both batteries, including anodes, cathodes, electrolytes and separators. In addition, the future research directions of Li–S and Li–O$_2$ batteries are also pointed out in this book, with several suggestions to solve the tough problems that limit the development of Li–S and Li–O$_2$ batteries, such as the lithium dendrites and electrolyte decomposition. This book also studies the potential applications of Li–S and Li–O$_2$ batteries, together with their challenges and perspectives discussed.

Dalian, China

Huamin Zhang
Xianfeng Li
Hongzhang Zhang

Contents

Li–S and Li–O$_2$ Batteries with High Specific Energy

Abstract This book introduces two important kinds of next-generation batteries: the lithium-sulfur battery and the lithium-air (or lithium-oxygen) battery. Both batteries have very high theoretical specific energy (2600 and 13,000 Wh/kg respectively), which could contribute to a longer use life of electronic devices. Based on the current available literatures, technical reports and data, a comprehensive review is made to introduce the basic principles, historical development, current status and future challenges of both battery technologies.

Keywords Li–S battery · Li–O$_2$ battery · High specific energy · Sulfur cathode · Lithium anode · Li-ion conductive electrolyte · Energy storage devices · Electric vehicles

1 General Introduction to Batteries with High Specific Energy

With the intensification of global environmental crisis, electric vehicles with the characteristics of zero emission significantly attract people's attention, leading to a rapid development throughout the world [1]. The key component of electric vehicles is the battery, which determines the endurance mileage of electric vehicles. Currently, the commonly used electric vehicle power supply is lithium-ion secondary batteries, and their specific energy is 100–200 Wh/kg, which could only allow a driving range of 160 km on a single charge based on the design of the ordinary electric car [2, 3]. The specific energy of lithium-ion secondary batteries could not meet the goal of the United States Advanced Battery Consortium (USABC) to achieve a driving range of 500 km on a single charge. As shown in Fig. 1, although lithium carbon fluoride and lithium thionyl chloride batteries have a very high specific energy (over 600 Wh/kg), they cannot be recharged, which limits their application in the field of electric vehicles. Besides, the specific energy of lithium-ion batteries is already close to its theoretical value, but it still falls short of the requirement of power supply for electric cars. Therefore developing other

© The Author(s) 2017
H. Zhang et al., *Li–S and Li–O$_2$ Batteries with High Specific Energy*,
SpringerBriefs in Green Chemistry for Sustainability,
DOI 10.1007/978-981-10-0746-0_1

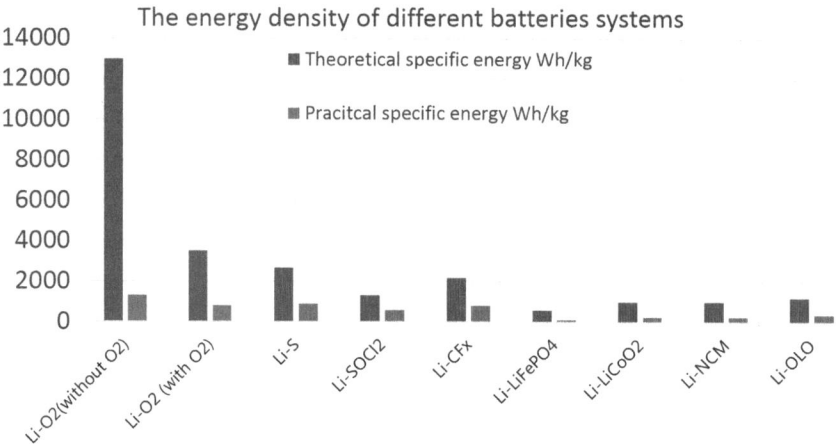

Fig. 1 The specific energy of different kinds of batteries

secondary battery systems with higher specific energy is urgent. Among the secondary power supply systems, Li–S battery owns a theoretical specific energy of 2600 Wh/kg, and Li–O₂ battery owns a theoretical specific energy of 3500 Wh/kg (based on Li_2O_2) or 13,000 Wh/kg (based on Li), which is much higher than the specific energy of lithium-ion battery (Fig. 1). Taking into account the ratio between the actual specific energy and the theoretical one, the practical specific energy of Li–S and Li–O₂ batteries is expected to reach 800 and 1700 Wh/kg, respectively. Therefore, Li–S and Li–O₂ batteries are considered as the next-generation battery technologies, with a very significant application prospect [4].

2 Research and Development of Li–S Batteries

2.1 General Introduction

In 1962, Herbert [5] filed the patent of using elemental sulfur as the battery cathode material, which is deemed as the origin of Li–S battery. In the 1980s, the lithium based batteries stepped into the industrialization stage. However, the cycle stability and safety of lithium-sulfur battery are very poor, due to the insulating nature of the sulfur and its compounds (S, polysulfide, and Li_2S), the shuttle effect of polysulfide (PS for short) [6–13], and the irreversible dissolution and deposition of lithium anode [14–27]. These problems were not solved effectively at that time, so the Li–S battery was not fully developed. In the 1990s, the research focus was shifted to the more stable sodium-sulfur battery system and lithium-ion battery system [28–37]. Until very recently, under the pressure of the energy and environmental crisis, the demand for batteries with higher specific energy became even more urgent than ever before, aiming to replace the traditional fossil energy in some areas such as vehicles. As a result, the Li–S batteries regained the high attention of human society.

Currently, most of the developed countries, including the United States, Japan, Russia and European Union, strongly support the development of Li–S battery technology. Japan's New Energy and Industrial Technology Development Organization (NEDO) planned to invest 30 billion yen (about 2.4 billion Chinese yuan) annually, and the goal is to develop Li–S batteries with a specific energy of 500 Wh/kg in 2020. The EU opened a "Horizon 2020" research and development program in 2015, planning to invest $7.6 million in electric vehicles using Li–S batteries. U.S. DOE has also invested a lot of manpower and resources to support the development of Li–S batteries. Meanwhile, the research and development of commercial Li–S battery has made significant progress in recent years worldwide, including companies like Sion Power (USA) [38], Polyplus (USA) [39] and Oxis Energy (UK) [40]. In 2010, Sion Power Company demonstrated the unmanned aerial vehicle (UAV) powered by solar cell in day time and Li–S battery (350 Wh/kg) in night time, creating a continuous flight record of 14 days [38]. Chemical Defense Institute of the Chinese People's Liberation Army has developed Li–S secondary batteries with a specific energy of 330 Wh/kg, which can maintain 60 % of its capacity after 100 cycles at 0.2 C. Dalian Institute of Chemical Physics developed Li–S primary batteries with a specific energy over 900 Wh/kg (1000 Wh/L) [41], which has already surpassed most traditional primary batteries. A Li–S battery can also produce a high specific power comparable to that provided by Ni–Cd batteries, which makes it very attractive for high energy-high power application. Unlike Ni–Cd batteries, Li–S batteries are not known to suffer from memory effect and are very tolerant to overcharging.

Currently, it is convinced that the technological advantages of Li–S battery are as follows:

1. Endow the mobile electronic devices with light weight and long use time after charging;
2. Meet the requirements of consumer electronics with low voltage power supply;
3. Easy to be designed as flexible and wearable devices;
4. Suitable for the electric vehicles and unmanned aircrafts;
5. Use the similar production line to that of lithium-ion batteries.

Even though, Li–S batteries need further breakthroughs in key materials and technologies in order to realize industrialization.

2.2 Principle and Clarification of Li–S Battery

Li–S battery, with sulfur or sulfur compounds as the cathode material and with lithium or lithium compounds as the anode materials, converts electrochemical energy and electric energy via the S–S bond cleavage and generation process. Similar to other Li-ion batteries, the Li^+ ions shuttle between anode and cathode of Li–S batteries during charge and discharge. The electrochemical reaction is as follows:

$$S + 2Li \underset{charge}{\overset{discharge}{\rightleftharpoons}} Li_2S, \Delta G = -425 \text{ Kj mol}^{-1} \tag{1}$$

Due to the lowest metal atom weight and the most negative potential of lithium, as well as the two-electron reaction of sulfur, the specific energy of Li–S battery is 3–5 times higher than that of traditional Li-ion or Ni–H batteries. The charge/discharge process of sulfur is quite complicated, with series of intermediate products coexisting in electrolyte and reacting with each other. As shown in Fig. 2, the long chain Li_2S_8 is mainly formed at high stage of charge (SOC), and the short chain Li_2S_2 and Li_2S is mainly formed at low SOC. Even though, nearly all the Li_2S_n (n ≤ 8) coexist in the Li–S battery, due to the chemical reaction between sulfur and lithium. Along with the in-depth understanding of sulfur-based chemical reaction, the performance of sulfur cathode has been improved significantly. For example, the utilization rate of sulfur has been improved from 50 % (about 800 mAh/g) to over 90 % (about 1500 mAh/g).

Compared with the currently used Li-ion insertion and extrusion materials, such as $LiCoO_2$, $LiFePO_4$, or $LiMn_2O_4$, the principle of sulfur species reaction is the dissolution and deposition of polysulfide at the interface between electrolyte and electrodes. The most commonly studied electrolyte material is composed of lithium salts and ether solvents, while the most widely used electrode material is carbon

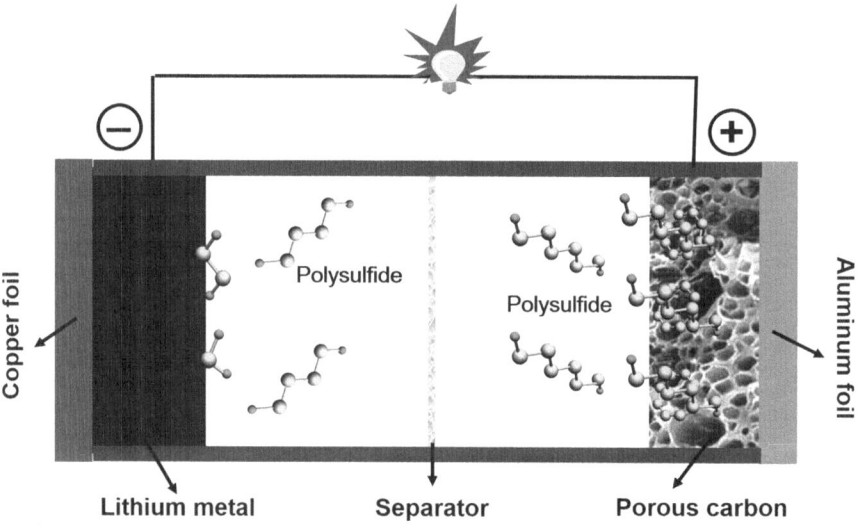

Fig. 2 The distribution of active materials in Li–S battery

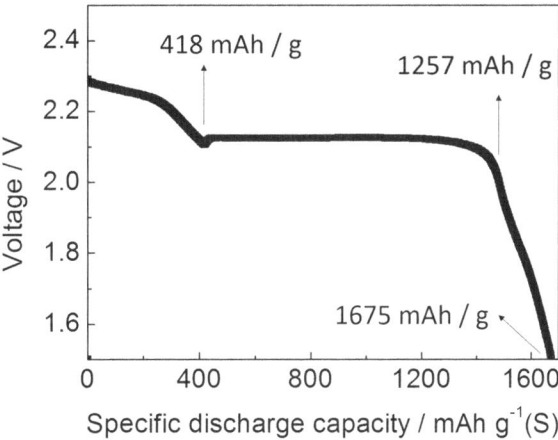

Fig. 3 The discharge stages of Li–S batteries

particles with a high specific surface area. As for the Li–S batteries, the C-rate capability is controlled by the PS diffusion in electrolyte. PS diffusion in Li–S batteries could be much faster than the Li^+ diffusion in crystals, which is the rate-determining step in the charge/discharge process of Li-ion batteries. As a result, it is possible to achieve higher specific power with Li–S battery than the Li-ion secondary battery (Fig. 3).

There are three stages during the discharge process of Li–S secondary batteries, as shown below:

(1) Reactions at the upper discharge plateau from 2.4 to 2.15 V (fast kinetics):

$$S_8^0 + 4e^- \rightarrow 2S_4^{2-}$$

(2) Reactions at the middle discharge plateau around 2.1 V (medium kinetics):

$$S_4^{2-} + 4e^- \rightarrow 2S^{2-} + S_2^{2-}$$

(3) Reactions at the lower discharge plateau from 2.1 to 1.5 V (slow kinetics):

$$S_2^{2-} + 2e^- \rightarrow 2S^{2-}$$

Besides the high specific energy and high specific power, Li–S batteries own some other potential advantages:

1. High and low temperature tolerance. Li–S battery has excellent performance in a wide temperature range from −40 to 80 °C, while it's difficult to charge the Li-ion battery at the temperature below −20 °C or above 80 °C.
2. Intrinsic safety mechanism. The phenomenon of lithium dendrite growth is not that serious in Li–S battery compared to other battery systems using lithium as the anode, due to the reaction between PS and lithium. However, this reaction

could accelerate the decomposition of electrolyte, and lead to the termination of the cycle life. This mechanism could help the Li–S batteries meet the safety requirement of practical application.

3. High power output. The rate-determining step in the discharge process of Li–S battery is the PS diffusion in electrolyte, which could be much faster than Li^+ insertion into crystal. As a result, the specific power of Li–S battery could reach a very high value, although it still has not been realized yet.

4. Simple control of the charge/discharge process. The Li-ion secondary battery needs a complicated charge program to avoid the risk of overcharging during the charge process and an extra electronic protection circuit at a high temperature of over 60 °C. However, the lithium-sulfur battery could endure the over-charge and over-discharge, which could be simply controlled.

5. Low cost of sulfur. The abundance of sulfur in the earth's crust ensures the low cost of Li–S batteries.

6. Compatible with the existing process for manufacturing Lithium-ion batteries. Since the configuration of Li–S secondary battery is similar to that of the lithium-ion batteries, the conventional lithium-ion battery production line can be directly used for Li–S batteries.

2.2.1 The Charge/Discharge Mechanism

Anode Mechanism

In the Li–S battery, lithium metal can be used not only as the current collector, but also as the active material. During discharging, the lithium metal is turned into lithium ions; during charge, the lithium ions are deposited on the lithium anode. There are many other secondary batteries using metallic lithium as the anode, such as $Li–O_2$ batteries and lithium metal based Li-ion batteries. As for different types of batteries, the chemical reactions and morphology changes of the lithium anode vary a lot during charge/discharge process. However, they commonly suffer from the "lithium dendrite" problem, which is occurring in the charge process. Although the lithium dendrite growth in Li–S batteries is much slower due to the reaction between lithium and polysulfide, it might still cause safety issues. Besides that, the lithium dendrite owns high reactivity, which could accelerate the electrolyte decomposition.

Cathode Mechanism

Because the elemental sulfur is insulating, it needs to be dispersed in conductive carbon materials to react. The Li–S battery is at the 100 % SOC after assembly. When the Li–S battery begins to discharge, the sulfur would turn into polysulfide (PS) and dissolve into electrolyte. At last, the PS will turn into Li_2S and Li_2S_2

deposited on the cathode and vice versa. Therefore, the sulfur cathode experiences two phase transitions during the charge/discharge process respectively. Due to the different density of sulfur (2.06 g cm^{-2}) and Li$_2$S (1.66 g/cm^3), their volume would shrink or expand during the charge/discharge process. Simultaneously, the particles of sulfur and lithium sulfide will depart from their original positions during the phase transition process, causing the decrease of capacity.

2.2.2 Polysulfide (PS) Shuttle Mechanism

What's PS Shuttle?

PS (Li$_2$S$_n$, 2 < n < 8) shuttle is one of the most serious problems for Li–S battery. Because PS is prone to be dissolved into ether electrolyte, during the charge process, the short chain PS is oxidized to long chain PS at the cathode, which would diffuse to the anode and form short chain PS with lithium. The short chain PS could diffuse back to the cathode to be oxidized to long chain PS again (shown as Fig. 4). The continuous PS diffusion between anode and cathode is named as PS shuttle effect.

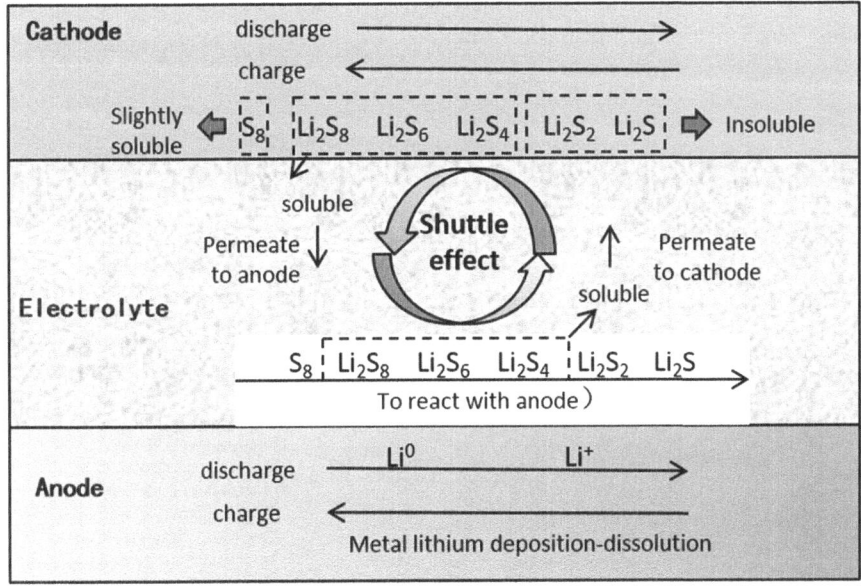

Fig. 4 The mechanism of PS shuttle in Li–S battery

Influence of Polysulfide Shuttle

The PS shuttle leads to two major problems. For one thing, the coulombic efficiency during the charge/discharge process will decrease to 90 % or even lower. For another thing, the lithium would react with the PS and electrolyte all the time, with the solid electrolyte interphase (SEI) keeping breaking and regenerating, which accelerates the electrolyte decomposition and leads to battery failure.

How to Deal with PS Shuttle?

In order to control the PS shuttle, the key materials of the Li–S battery have to be further developed. These materials include sulfur hosts, electrolytes, separators and novel anodes. The basic shuttle inhibition mechanisms of different battery components are shown in Fig. 5.

2.2.3 Capacity-Fading Mechanism

The capacity fading of Li–S battery during cycling is another key issue that needs to be solved. In general, the capacity fading phenomenon is related to three components of Li–S battery: the anode, the electrolyte and the cathode. Their capacity fading mechanisms are illustrated as below.

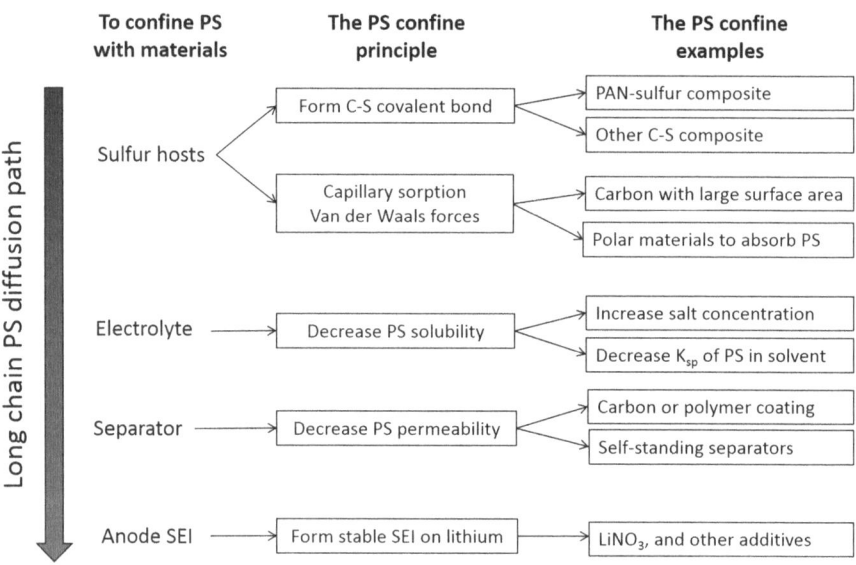

Fig. 5 The PS shuttle inhibition mechanisms of different battery components

Anode Capacity Fading

As illustrated in Sect. 2.2.1, the lithium metal is continually dissolved and deposited on the anode during the discharge and charge process. During charge, the lithium particles might be unevenly deposited on the anode, which look like dendrites and these dendrites would be shed off during cycling. As a result, the active lithium will be consumed as dead lithium, which leads to the decrease of battery capacity. The mechanism could be described in Fig. 6.

In fact, the lithium batteries are always designed to have an excess amount of lithium if using lithium and sulfur as active materials. In that case, the lithium will not cause capacity fading in the first several cycles. However, if the excess lithium is used up, the capacity of the battery decays quickly.

Cathode Capacity Fading

The capacity fading of cathode is mainly caused by the aggregation of charge and discharge solid products (Li_2S, Li_2S_2 and S_8), as shown in Fig. 7, which leads to the worse contact between active materials and electrodes. Besides that, the aggregated particles of Li_2S and S could block the Li^+ transport channels in the cathode, which hampers the subsequent reactions. In addition, a portion of PS would diffuse to the

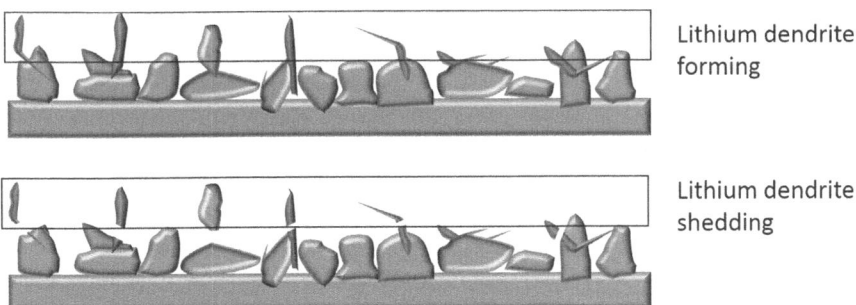

Lithium dendrite forming

Lithium dendrite shedding

Fig. 6 Anode capacity fading mechanism of Li–S battery

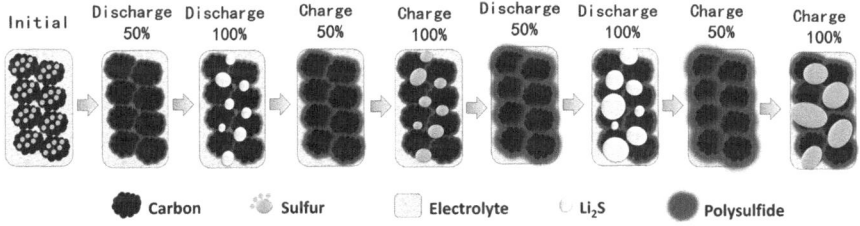

Fig. 7 The capacity fading mechanism of Li–S cathode

anode and the separator, which can not get back to the cathode, leading to the capacity decay. Because the latter part of capacity loss is relatively constant, accounting for about 10–20 % of the total capacity, the aggregation of charge and discharge products is the major reason for the capacity decay of Li–S cathode.

Capacity Fading Caused by Electrolyte

During the charge/discharge process, the electrolyte keeps reacting with the lithium anode and forms SEI. Meanwhile the electrolyte would be consumed, leading to the increase of PS concentration. As a result, it becomes more difficult for the PS to diffuse at the cathode/electrolyte interface, causing heavy concentration polarization, and then the charge/discharge process is terminated prematurely at the cut-off voltage. Finally, the electrolyte would dry out and the Li–S battery would never be charged or discharged again. For another thing, during the electrolyte decomposition, some gases such as H$_2$, CH$_4$ and CO could be generated, which further disturbs the continuity of electrolyte and decreases the stability of the battery. And even worse, the package of battery might be split up due to the excessively high inner pressure.

2.2.4 Battery Configuration

The battery performance of Li–S battery is affected significantly by its configuration, such as the choice of key materials, the uniformity of electrodes, the assembly pressure, the rigidity of separators and the amount of lithium and electrolyte [42]. Based on common sense, using excess lithium and electrolyte is beneficial to the cycle stability of Li–S battery. However, because it would pull down the specific energy of Li–S battery, the amount of electrolyte should be controlled to be less than 5 folds of the amount of sulfur and the amount of lithium should better be 1.5–2.0 folds of the theoretical value. In this situation, it is of great importance to assemble the battery with suitable pressure between anode and cathode, which could help decrease the amount of lithium dendrite.

Coin Type Cell

Coin type cell is the most commonly used cell for Li–S battery research, as well as for Li-ion batteries, due to its simple structure and assembly process. In this type of battery, the cathode, the separator and anode are pressed together tightly. However, the specific energy of this kind of cell is very low, because the steel shell is extremely heavy (Fig. 8).

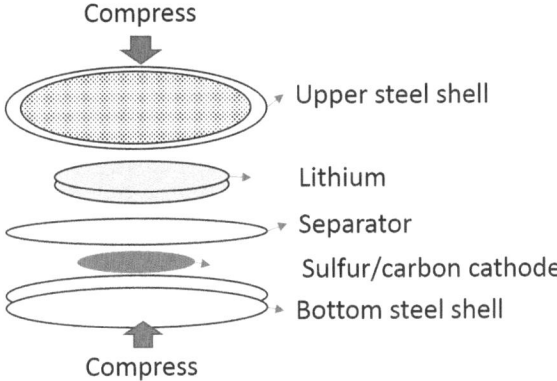

Fig. 8 The structure of the coin type Li–S battery

Pouch Cell

The most applicable Li–S batteries are assembled in the pouch cells, which have a structure similar to that of the Li-ion batteries. The cathode and anode could be welded or stacked together, with separator and electrolyte between them. The pressure on the surface of the pouch cell is applied by vacuum sealing or external equipment (Fig. 9).

2.3 Research Status of Li–S Batteries

Currently, the research of Li–S battery still focuses on the development of novel materials for anode, cathode, electrolyte and separator, in order to achieve satisfactory battery performance.

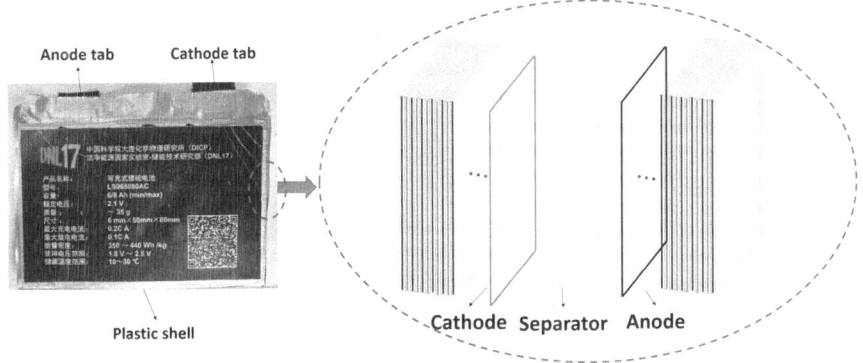

Fig. 9 The structure of the soft package Li–S battery

2.3.1 Electrolyte

The electrolyte in Li–S battery plays two roles: to transport Li^+ between cathode and anode; to transport PS at the cathode/electrolyte interface. As the PS shuttle could cause the consumption of electrolyte and lithium, as well as low coulombic efficiency, the PS diffusion in electrolyte must be well confined. The ideal electrolyte for Li–S battery should possess the following characteristics: (1) excellent chemical and electrochemical stability, and it should not decompose or react with other battery materials; (2) excellent Li-ion transport ability; (3) insulating; (4) low-cost; (5) environmentally friendly. The currently developed battery systems could be divided into three categories according to different types of electrolyte materials: the liquid battery system, the solid battery system and the semi-flow battery system.

The Liquid Electrolyte

As for the liquid Li–S battery, the liquid electrolyte solution is composed of solvents, salts and functional additives. At the beginning, the solvent of Li–S battery electrolyte was made from DMSO, DMF and THF, then it gradually developed to aliphatic amines, lipids, ether solvents [43–55] and block copolymers [56]. In 1989, Peled et al. proposed that 1,3-dioxolane (DOL) could be used as the solvent of Li–S battery electrolyte. The DOL owns low viscosity and high lithium solubility, and could form a protective layer on the anode to enhance cycle life. However, the discharge products of sulfur are prone to form Li_2S_2 but not Li_2S, leading to about 50 % capacity loss [57]. As a result, the DOL is usually mixed with DME, to achieve better comprehensive performance [58]. The currently applied electrolyte is composed of DOL and DME (with volume from 1:2 to 2:1), with 1 M LiTFSI salt added. In order to control the PS shuttle, Mikhaylik [59] proposed to add "N–O" composites into the electrolyte, and $LiNO_3$ is found to be the best choice. The $LiNO_3$ could react directly with lithium, electrolyte and PS to form a dense SEI composed of Li_2O, Li_3N, $LiSO_3$ and etc., which could stop the further reaction. However, the $LiNO_3$ could be irreversibly consumed at the cathode surface below 1.7 V, which would occupy the active sites for sulfur reaction. Besides that, the $LiNO_3$ is a strong oxidant, which is dangerous if blended with sulfur and carbon. The mixture has a composition similar to that of the black powder which is one of the four great inventions in ancient China. In addition, some scholars proposed that the LiBOB and P_2O_5 could be used to protect the lithium anode surface, with a layer of stable SEI [60, 61]. According to the theory of solubility, the saturation concentration of PS could be effectively reduced with the increase of Li^+ concentration [62]. However, once the lithium salt concentration is increased to over 4 M, the viscosity and density of the electrolyte would be too large, hindering its use. Besides the electrolytes mentioned above, the ionic liquids such as [PP14] [TFSI] (1 M LiTFSI) and PP13-TFSI [63, 64] could also be used to enhance the safety of electrolyte. However, the viscosity and density of ionic liquids should be further decreased in order to improve the capacity delivery of sulfur [42]. Currently, there still exists a paradox in the development of electrolytes: on one hand, the PS diffusion should be confined to prevent the chemical reaction between PS and lithium; on the other hand, the PS diffusion should be enhanced to increase the capacity delivery of sulfur. This paradox is difficult to resolve and affects the development of Li–S batteries.

The Solid Electrolyte

The solid electrolyte could avoid the battery burning and electrolyte leakage, decrease the PS shuttle, and inhibit the lithium dendrite growth on the anode. It could be divided into two series, one is the gel electrolyte composed of polymers and lithium salts, and the other is the pure inorganic Li^+ conductor [65–69]. Gel electrolyte is obtained by adding plastic and lithium salt into the polymers, such as PAN, PMMA, PEO, PVA, PVDF and their derivatives. The ionic conductivity of the gel electrolyte is between that of liquid electrolyte and inorganic electrolyte [70]. The most popular gel electrolyte is composed of PEO and lithium salt, such as $LiASF_6$ or LiTFSI. However, this kind of electrolyte owns very low lithium-ion conductivity, so it has to be used over 60 °C [71–73]. Hassoun et al. [64] used the NCPE ($PEO/LiCF_3SO_3^+$ 10 % ZrO_2) to assemble $Li/NCPE/S$–C and $Li/NCPE/Li_2S$–C all-solid-state batteries, which obtained 100 % coulombic efficiency at 70 and 90 °C respectively, with no PS shuttle effect. The inorganic electrolyte is mainly composed of the fast Li-ion conductor, which is cut into thin slices from bulk or pressed into membranes using crystal and amorphous powder. This kind of electrolyte could stop the diffusion of PS to anode and has great potential to solve the PS shuttle problem [16, 74–78], however, the solid electrolyte still faces the problems such as high interface resistance, and it is too creepy to achieve large-scale production [70, 79].

2.3.2 Anode Materials

The research and development of lithium anode material started in the 1950s, when the deposition of lithium was observed in organic electrolyte. After 20 years, the lithium anode was successfully applied to the primary batteries such as Li–CF_x, Li–MnO_2 and Li–$SOCl_2$, which have been commercialized for a long time. However, the application of lithium anode in the secondary batteries, such as Li–TiS_2, Li–MnO_2, Li–MoS_2 has suffered from serious problems, such as the lithium dendrite, lithium mossy and electrolyte decomposition. The anode materials for Li–S batteries developed up to now are listed as follows:

1. Lithium metal

Due to the high reactivity of lithium metal, its behavior in battery is strongly depending on its surface composition. The currently used electrolyte in lithium based batteries is composed of lithium salt and organic electrolyte. Because most of the lithium salts and organic solvents are unstable on lithium anode surface, they would react with each other and form a passivation layer of organic and inorganic compounds at the electrolyte/lithium interface.

The passivation layer formed on the lithium anode surface owns properties distinctly different from the lithium metal, but similar to that of the solid state electrolyte. In 1979, Israel scholar Emanuel Peled named the passivation layer as Solid Electrolyte Interphase (SEI). The SEI could effectively inhibit the permeation of electrolyte and stop the reactions between electrolyte and lithium. After nearly 30 years of research, the formation mechanism and influence of SEI on battery performance have been gradually clarified.

The SEI composition on lithium electrode was fully characterized by spectroscopy, photoelectron spectroscopy and NMR techniques. The major composition of SEI in different solvents and salts is shown in Table 1.

Accompanying the SEI forming process, some gases (H$_2$, CH$_4$, CO), as the decomposition products of electrolytes, are also generated continuously at the lithium anode surface during cycling. As a result, the inner pressure of the Li–S battery would keep increasing and cause safety issues.

In addition, the SEI on the lithium anode is so creepy that is prone to break down during the lithium dissolution and deposition process. The SEI forming and

Table 1 The major composition of SEI on lithium electrode in different electrolytes

Electrolyte	SEI composition	Reaction product on lithium anode
Alkyl carbonates	Propylene carbonate (PC)	CH$_3$CH$_2$OCO$_2$Li, Li$_2$CO$_3$, LiOH
	Ethylene carbonate (EC)	(CH$_2$OCO$_2$Li)$_2$, CH$_3$CO$_3$Li, Li$_2$CO$_3$
	Dimethyl carbonate (DMC)	CH$_3$OCO$_2$Li, Li$_2$CO$_3$
	Diethyl carbonate (DEC)	CH$_3$CH$_2$OCO$_2$Li, CH$_3$CH$_2$OLi
	Ethyl methyl carbonate (EMC)	CH$_3$OLi, CH$_3$OCO$_2$Li, LiOH, Li$_2$O
Other esters	Methyl formate (MF)	HCO$_2$Li, CH$_3$OLi
	γ-Butyrolactone (γ-BL)	CH$_3$(CH$_2$)$_2$CO$_2$Li, LiO(CH$_2$)$_3$COOLi
Ethers	Tetrahydrofuran (THF)	CH$_3$(CH$_2$)$_3$OLi, LiOH
	Methyltetrahydrofuran (2Me-THF)	ROLi, Li$_2$O
		ROLi(CH$_3$OLi)
	Ethylene glycol dimethyl ether (DME)	CH$_3$CH$_2$OLi, (CH$_2$OLi)$_2$, (OCH$_3$CH$_2$OCH$_2$)n
	Dioxolane (DOL)	CH$_3$OLi, CH$_3$OCH$_2$CH$_2$OLi,
	Diglyme (DG)	(CH$_2$OLi)$_2$
Blend solvents	EC-PC	Similar to the products with each solvent
	EC-DEC	
	MF-EC	
	MF-DMC	
Lithium salt	LiAsF$_6$	LiF, LiAsF$_y$
	LiClO$_4$	Li$_2$O, LiCl, LiClO$_3$, LiClO$_2$
	LiBF$_4$	LiF, Li$_x$BF$_y$, LiBF$_y$O$_z$
	LiPF$_6$	LiF, Li$_x$PF$_y$, LiPF$_y$O$_z$
	LiSO$_3$CF$_3$	Li$_x$S$_y$O$_z$, LiF, RCF$_y$Li$_z$
	LiN(SO$_2$CF$_3$)$_2$	LiF, LiS$_y$O$_z$, Li$_3$N, RCF$_y$Li$_z$, LiNSO$_2$CF$_3$

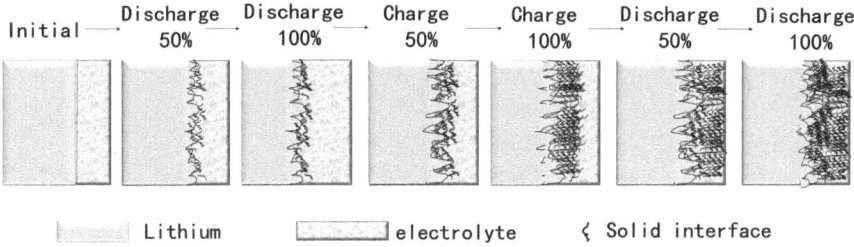

Fig. 10 The morphology change of SEI during charge and discharge in liquid electrolyte

breaking process will take place all the time, and finally the electrolyte would dry out. Besides that, the broken SEI would accumulate loosely on the lithium anode surface, making the battery volume swell gradually. It would break the shell of the battery, as well as break the shell of the electric devices. This is a big problem for using liquid electrolyte to contact directly with lithium anode (Fig. 10).

In order to solve above problems, researchers did a lot of work on the selection of suitable electrolytes, including many kinds of solvents, lithium salts and additives, in order to improve the stability of the SEI. Firstly, suitable solvents should be chosen. It's found that ether shows better interface compatibility with lithium metal and higher lithium cycling efficiency compared to ester. And the F substituted solvent might be better, because it could promote the formation of a LiF protective layer on lithium anode [80]. Secondly, suitable lithium salt should be chosen. LiTFSI was found to effectively form a LiF layer to protect the lithium anode surface, and other lithium salts such as $LiPF_6$, $LiAsF_6$ and LiFSI also own the similar function. Increasing the concentration of lithium salt was also proved to be effective to protect the lithium anode surface. Thirdly, suitable additives should be chosen. For example, HF [81], CO_2 [82] could change the composition of SEI; SnI_2 and AlI_3 [83] could achieve a uniform potential distribution on the lithium anode surface; Cs^+, Ru^+ [84] could reduce the formation of lithium dendrites; $LiClO_4$ [6] and $LiNO_3$ [11, 52, 85–88] could form chemically stable SEI. According to literature, the $LiNO_3$ is the most effective additive, which could achieve 100 % coulombic efficiency. Although the safety and cycle stability could be enhanced by these methods, the lithium metal is still far from being used in large-scale commercial applications.

2. Lithium alloys

In order to inhibit the electrolyte decomposition, Li based alloy was also proposed to improve the cycle life of Li–S battery. Alloys such as Li–Si, Li–Al and Li–Mg were proposed to improve the performance of Li–S batteries, however, it did not show obvious improvement.

3. Li^+ insertion/extrusion anode materials

Given the problems of lithium metal anode, researchers proposed to use other anode materials to substitute the lithium metal. Currently developed anode materials could

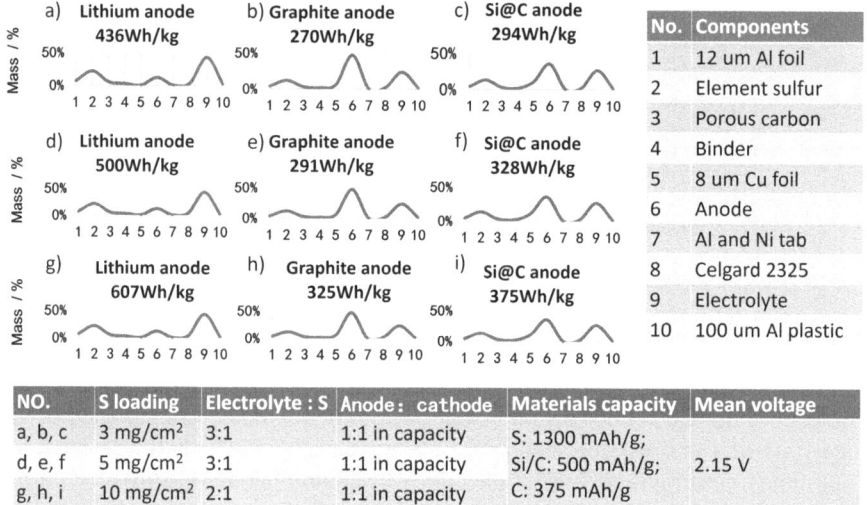

No.	Components
1	12 um Al foil
2	Element sulfur
3	Porous carbon
4	Binder
5	8 um Cu foil
6	Anode
7	Al and Ni tab
8	Celgard 2325
9	Electrolyte
10	100 um Al plastic

NO.	S loading	Electrolyte : S	Anode: cathode	Materials capacity	Mean voltage
a, b, c	3 mg/cm²	3:1	1:1 in capacity	S: 1300 mAh/g;	
d, e, f	5 mg/cm²	3:1	1:1 in capacity	Si/C: 500 mAh/g;	2.15 V
g, h, i	10 mg/cm²	2:1	1:1 in capacity	C: 375 mAh/g	

Fig. 11 Practical specific energy of Li–S batteries with Li, C and C/Si anodes, based on calculation

be mainly divided into Si group, C group, C/Si composite group and other anode materials, which have advantages and disadvantages respectively. For example, although the theoretical capacity of silicon is 4200 mAh/g, its reversibility is very poor. And although the commercial C and C/Si composite own good reversibility, their theoretical capacity is too low (always below 500 mAh/g). Based on the calculation of all the battery components, the Li–S battery with Si/C composite anode could only achieve a specific energy of about 370 Wh/kg, which is much lower than that of metallic lithium anode (about 600 Wh/kg). The major components and their mass percent composition are shown in Fig. 11. There is no doubt that the lithium anode in Li–S batteries can achieve a super high specific energy. However, due to the excellent cycling stability of the Li$^+$ insertion/extrusion anode materials, they are also of great significance in the development of better anode materials. In addition, because these materials lack lithium, Li$_2$S has to be used as the cathode. But because the Li$_2$S is sensitive to water and oxygen, it is difficult to realize large-scale production.

2.3.3 Cathode Materials

At the end of the 20th century, people already had a clear understanding of the working mechanism of sulfur cathode, however, there is a lack of effective methods to improve the C-rate capacity and cycle stability. During the discharge/charge process, the S and Li$_2$S transform into each other, with the intermediate product polysulfide (PS). Because the PS (Li$_2$S$_n$, n = 3–8) would be dissolved in the electrolyte, it is hard to keep the original distribution of sulfur on the cathode during cycling, leading to the capacity decay of cathode. In addition, the PS shuttle effect would occur. The key task is to stop the diffusion of PS from cathode to anode and vice versa. So, great effort has been engaged in inhibiting the dissolution of PS to increase the cycle stability of Li–S battery. However, from another point of view, the dissolution of PS is necessary for the capacity delivery of sulfur, because the sulfur species [such as S, Li_2S and Li_2S_n (n = 3–8)] are all insulators and the reactions have to occur at the electrolyte/electrode interface. If the PS is insoluble in electrolyte, it would cover the surface of the carbon cathode and stop the electron transport. From this view point, it's the dissolution of PS that promotes the capacity delivery of sulfur species.

In recent years, the rapid development of nanotechnology has brought new opportunities for developing high-performance sulfur cathode with a variety of micro-and nanostructures. The sulfur could be dispersed in the micro/nano porous structures, thereby facilitating e^-/Li^+ conduction and improving the rate capability. It is discovered that the micro pores (<2 nm) and meso pores (2–50 nm) could inhibit the PS diffusion from cathode to anode. Due to its enormous porous structures, carbon materials are the most commonly used host materials for sulfur. Ever since Linda F. Nazar group [89] reported CMK-3 as the sulfur hosts (Fig. 12), a lot of carbon materials with different pore sizes and morphologies were developed, including nanotube, nano-fiber, graphene, nano flower and etc.

As reported by Guo et al. [90–92], the diameter of S_8 molecule is about 0.68 nm, which could fill into the carbon pores via melt diffusion or vapor deposition (Fig. 13). The PS could be confined in the carbon hosts by capillary sorption, and the confinement effect will be diminished with the increase of pore size, in the order of micro pore > meso pore > large pore. However, the meso pores and large pores are urgently needed for the sulfur accommodation. While for C-rate capability and capacity delivery, the ordered porous structure is much better than the disordered one. All in all, the carbon hosts with ordered hierarchically porous structure could achieve better performance.

Due to the non-polar characteristics of the carbon surface and the strong polarity of PS, the van der Waals force between them is quite weak, and the PS confinement effect is not ideal enough in pure carbon host. So researchers are prone to add some hetero atoms (O, N, S) on the carbon surface to increase its polarity [93–95], which however at the cost of decreasing the electronic conductivity and capacity delivery. For example, the g-C_3N_4 (Fig. 14) [96] owns excellent cycle stability but only has a specific capacity of 900 mAh/g, which is 3/4 of that of the commercial KB 600 particles [96].

Some researchers also tried to directly graft sulfur onto other molecules, such as the PAN-S or Poly(S-r-DIB) co-polymer (Fig. 15), by opening the S_8 ring at high temperature to improve the cycle stability [97–99]. However, the sulfur content of these materials is always low, which could not achieve high specific energy of Li–S battery.

Besides that, large amounts of conductive macromolecules, inorganic materials and polymers have also been used to prevent the diffusion of PS, by forming a layer

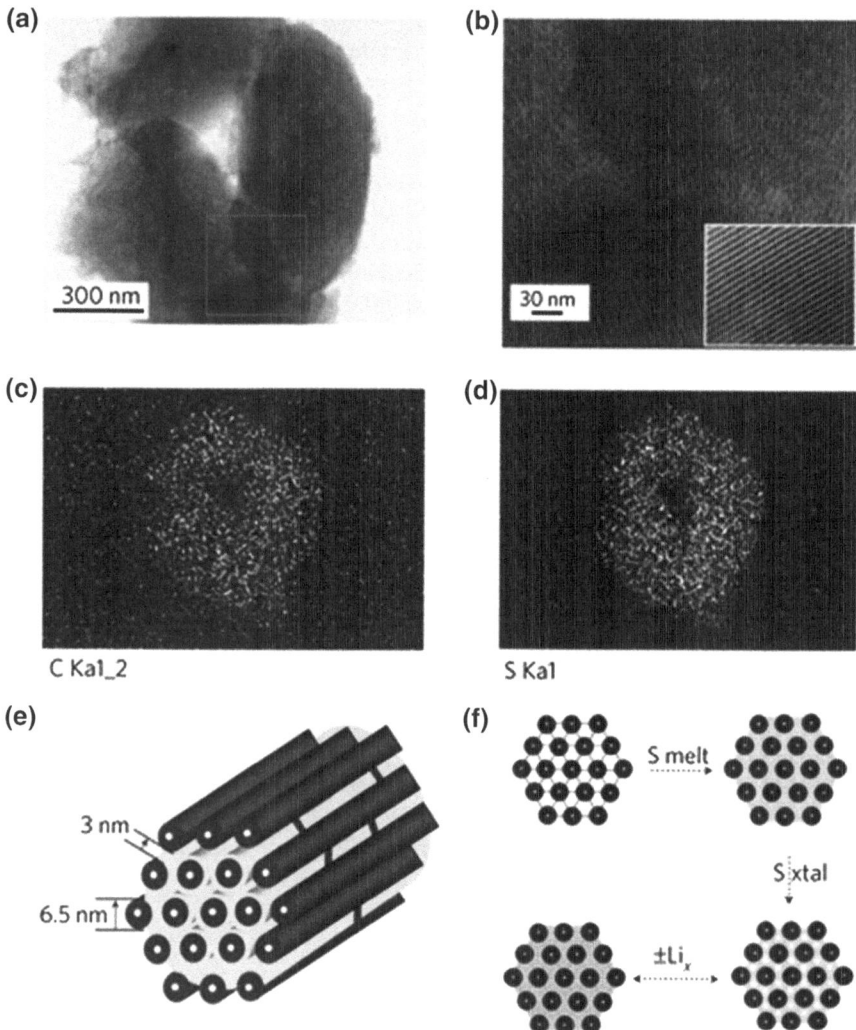

Fig. 12 **a** CMK-3/S-155 composite particle. **b** Image expansion corresponding to the area outlined by the *red square* in **a**, where the inset shows the TEM image for pristine CMK-3 at the same magnification. **c**, **d** Corresponding carbon and sulphur elemental maps showing the homogeneous distribution of sulphur. **e** A schematic diagram of the sulphur (*yellow*) confined in the interconnected pore structure of mesoporous carbon, CMK-3, formed from carbon tubes that are propped apart by carbon nanofibres. **f** Schematic diagram of composite synthesis by impregnation of molten sulphur, followed by its densification on crystallization. The lower diagram represents subsequent discharging–charging with Li, illustrating the strategy of pore-filling to tune for volume expansion/contraction. Reprinted with the permission from Ref. [89]. Copyright 2009 Nature Publication Group

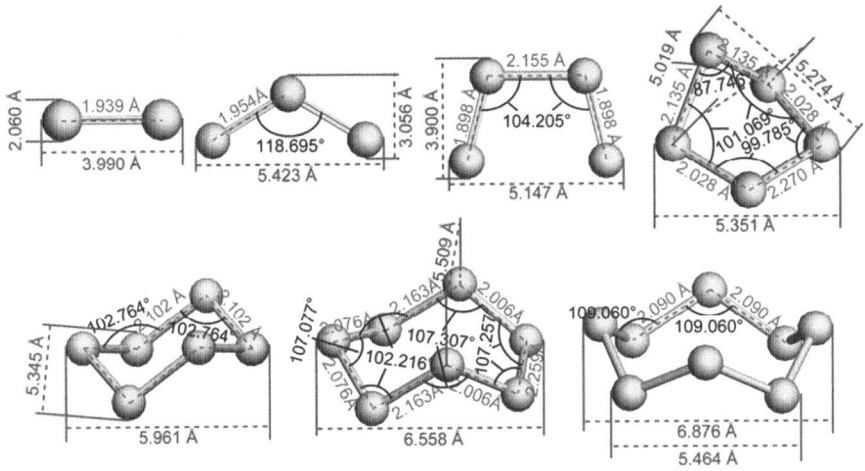

Fig. 13 Micro structure of sulfur species. Reprinted with the permission from Ref. [92]. Copyright 2013 John Wiley and Sons

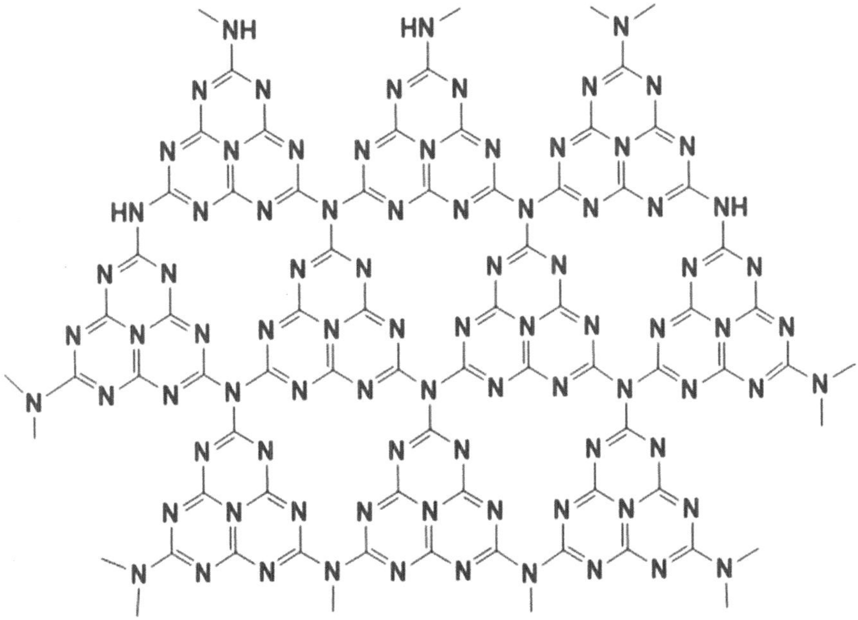

Fig. 14 Illustration of the molecular structure of g-C_3N_4. Reprinted with the permission from Ref. [96]. Copyright 2015 American Chemical Society

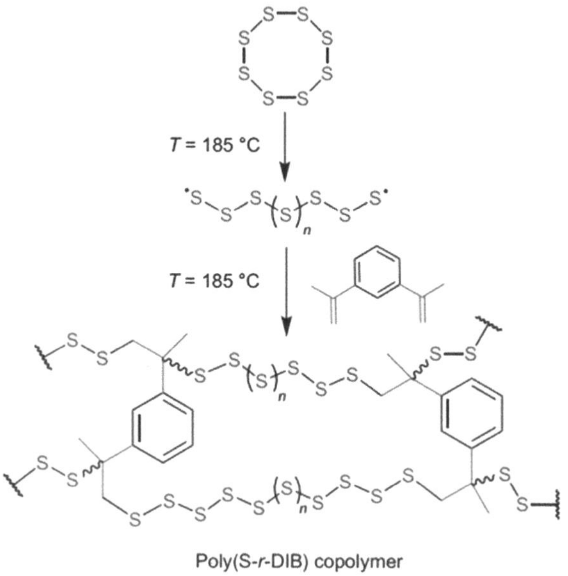

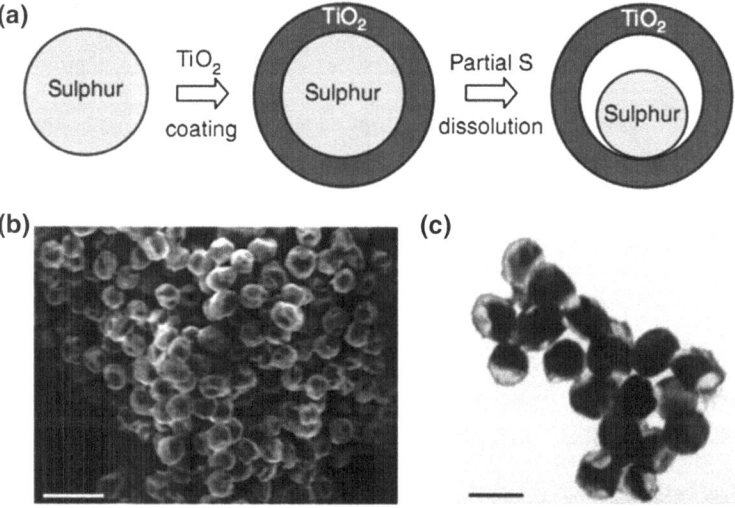

Poly(S-r-DIB) copolymer

Fig. 15 The molecule structure of the sulfur grafted onto Poly (S-r-DIB). Reprinted with the permission from Ref. [97]. Copyright 2013 Nature Publishing Group

of barrier on sulfur. These materials including polyaniline, polythiophene, TiO₂, etc. [100, 101]. For example, the TiO₂ and polyaniline could cover the sulfur with a yolk-shell structure (Figs. 16 and 17), to prevent the polysulfide diffusion and the volume expansion from S to Li₂S. However, the electronic conductivity of these

Fig. 16 The TiO₂/S composite with a yolk-shell structure

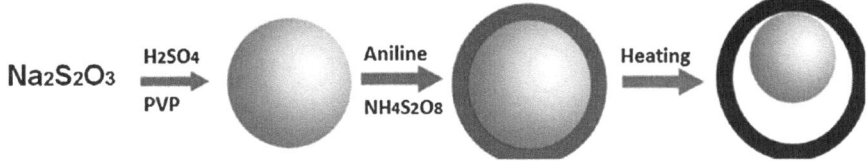

Fig. 17 The polyaniline/S composite with a yolk-shell structure. Reprinted with the permission from Ref. [101]. Copyright 2013 American Chemical Society

materials is relatively lower than the carbon materials, which needs further improvement.

Based on the statistics, half of the published research work is related to cathode materials. The roadmap of cathode material development could be illustrated in Fig. 18, from the simple blend of carbon and sulfur to the fixation of sulfur with carbon pores and chemical bonds. And the same time, the PS confinement mechanism has been gradually understood. Based on the practical demands at present, filling sulfur into the hierarchical carbon hosts is of great commercial significance. Besides that, the structure of the binder materials [102] also affects the battery performance a lot.

Currently, the cathode materials play an important role in the capacity delivery of sulfur and the PS confinement, the inherent problems of electrolyte, separators and anode materials still need further research work.

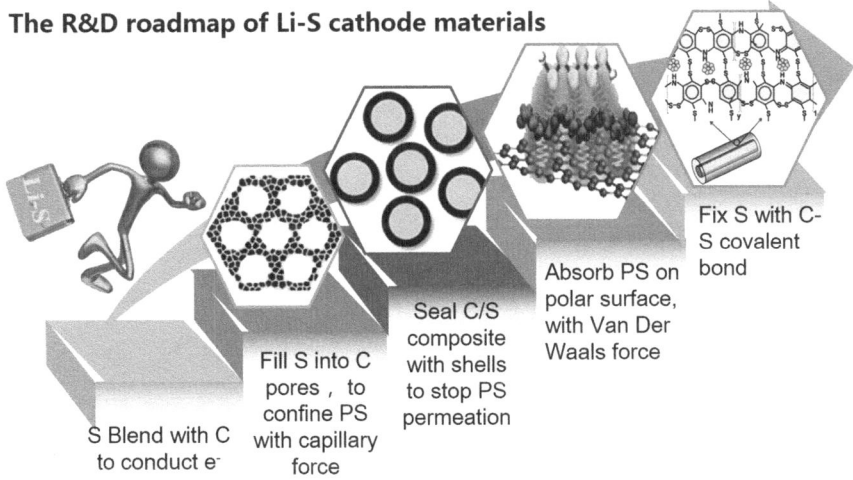

Fig. 18 The R&D roadmap of cathode materials for Li–S battery

2.3.4 Battery Separators

The practical use of lithium-sulfur battery separator needs to meet the following requirements: (1) high lithium-ion conductivity and low polysulfide ion permeability; (2) good mechanical and chemical stability to prevent the electrochemical corrosion and puncture of lithium dendrites; (3) low cost, allowing large-scale production. The currently used separators for Li–S battery are the same as Li-ion batteries, including the microporous polymer membranes, inorganic ceramic membranes, and all solid polymer electrolyte membranes [54]. The microporous polymer membranes, such as the PP or PE based ones, possess pores larger than 30 nm and could not stop the diffusion of S_n^{2-} (n > 2). Recent research shows that some membranes could help to confine the polysulfide shuttle [7, 13, 103–118]. For example, membranes made from PEO, V_2O_5 and Li_2S-P_2S_5 ceramic could separate the polysulfide by its dense structure while conduct the Li^+ via its –c–o–c– chain or unique crystal structure [110, 119]. What's more, gel membranes like micro-porous PVDF/Li-TFSI could also relief the PS migration, partially due to the higher mass diffusion resistance through the gel electrolyte soaked in the membrane [111, 120, 121]. Besides that, some interlayers made from carbon materials (such as CNT, Super P, graphene, reduced graphene) or nickel foam foil also help to trap polysulfide via surface adsorption or chemical bonding [112–118]. In addition, the inorganic ceramic membranes are too creepy for large-scale application, although they show excellent performance in separating the Li^+ and S_n^{2-} [122].

In recent years, the sulfonic acid type ion exchange membrane has been proposed for application in lithium-sulfur battery by the Chinese National Defense University, Tsinghua University and other research institutes [7, 123–126]. The sulfonic acid group ($-SO_3^-$) with negative charges is utilized to separate the polysulfide ion (S_n^{2-}) via the "charge repulsion" effect (Fig. 19). The study found that the transport number of lithium ion through this membrane is close to 1.0, showing excellent performance in separating lithium polysulfide. However, the Li^+ conductivity of this membrane is much too low (10^{-5} S cm^{-1}), which needs further improvement.

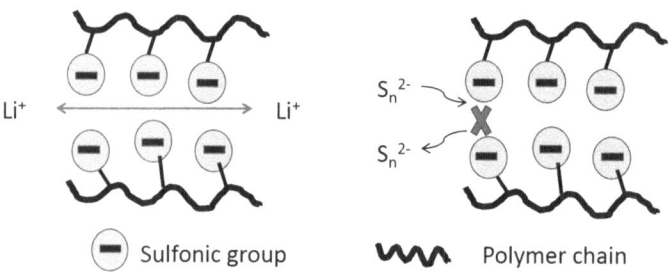

Fig. 19 The charge repulsion effect to separate S_n^{2-} and Li^+

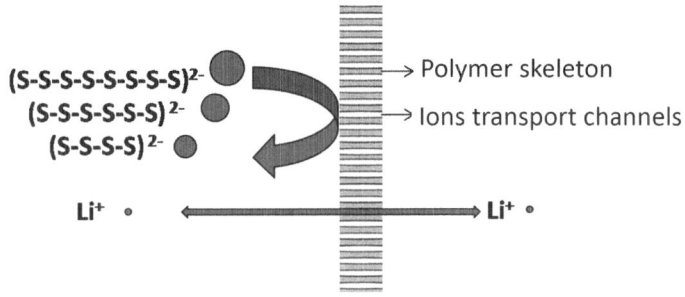

Fig. 20 The illustration of separating polysulfide ions and Li$^+$ by size sieving effect

As it is known, the soluble polysulfide is constituted by 4–8 sulfur atoms (diameter is about 0.5 nm), which is much larger than the single Li$^+$ (diameter is about 0.076 nm) as calculated by a DFT method [90, 127, 128]. It is possible to separate the polysulfide and Li$^+$ by the size sieving effect [129, 130], such as the nano-Li$^+$-channel PVDF polymer separators and the metal–organic-framework (MOF) based separators (Fig. 20).

2.3.5 Binder Materials of Cathode

Because the binder materials are always insulating, it should be better to have less binders on the cathode, with the purpose of ensuring high electronic conductivity and specific energy of the whole electrode. There are several binder materials applied to the development of C/S composite cathodes, such as polyvinylidene fluoride (PVDF), polytetrafluoroethylene, cellulose, protein, chitosan, polyethylene glycol, polyvinylpyrrolidone, and etc. It's found that the binding strength of PVDF is not good enough for Li–S cathode application, although it's commonly used for the Li-ion batteries, because the C/S composite is always too small in size and has too large specific surface area to be bonded onto the current collector (usually the aluminum foil). On the contrary, the water-soluble binder materials always have better binding ability, such as the alginate and cyclodextrin, because they own lots of hydrophilic functional groups, which could enhance the contact with the aluminum foil. By using the cross-linkable binders, such as PVP and Nafion shown in Fig. 21 [131], the binder content could be decreased to 0.5 wt% of the whole electrode, while keeping better binding strength compared to the cathode with 10 wt% PVDF. Besides that, a lot of binder-free cathodes were also developed. For example, sulfur is coated on the 3-D carbon networks, such as graphene, carbon nano fiber, carbon cloth, carbon paper, etc. However, based on the current results, the binder-free materials are difficult to be utilized in practical applications.

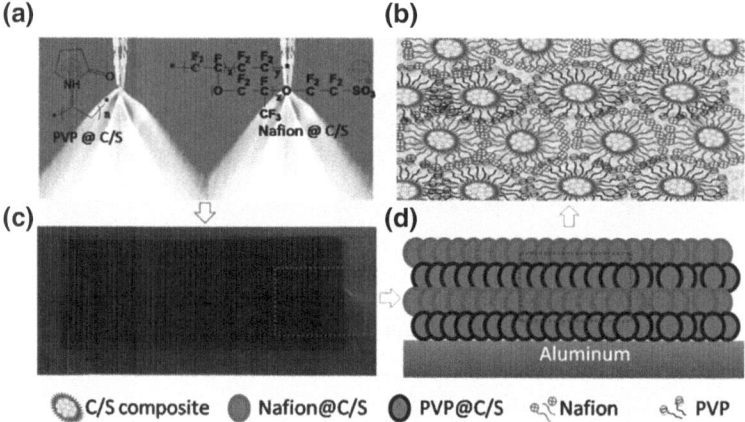

Fig. 21 The binding mechanism of Nafion and PVP for Li–S battery application. Reprinted with the permission from Ref. [131]. Copyright 2015 from American Chemical Society

2.4 Further Research Directions of Li–S Batteries

To achieve the practical application, the research should be focused on improving the overall performance of Li–S batteries, including the specific energy, specific power, energy efficiency, cycle life, shelf life, safety and so on. Since the practical applications in different fields have very different performance requirements for Li–S batteries, the research directions of Li–S batteries should be different as well. Some general research directions of Li–S batteries are listed as follows:

1. Develop cathode material with excellent comprehensive performance, including efficient utilization of sulfur, large charge/discharge rate, strong confinement of sulfur and moderate amount of liquid absorption, to ensure the Li–S battery has a specific energy of not less than 300 Wh/kg. Besides that, novel catalysts could be developed to accelerate the cathode reaction in Li–S battery, or change the reaction process of cathode to avoid PS diffusion.

2. Design the micro-nano structure of whole cathode, which is not limited to the C/S composites, for optimizing the mass transport and PS diffusion according to the requirements of practical applications. Firstly, the sulfur content and loading on the cathode should be high enough; secondly, the electrolyte distribution (or Li$^+$ transport) across the cathode should be optimized from macro to micro scale; thirdly, the electronic conductivity through the whole cathode should be ensured. Besides that, the PS diffusion could be confined by tuning the sulfur and carbon distribution in the whole cathode.

3. Study the safety of Li–S batteries with both experimental and simulation methods. Especially the safety of Li–S batteries under high (over 60 °C) and low temperatures (below 0 °C), when the electrolyte decomposition and lithium dendrite formation are prone to occur.

4. Develop high-performance anode material, including lithium metal, lithium alloy, Li^+ insertion material (silica, carbon, and etc.) and transfer material (metal oxide, etc.) [132–134]. The growth and shedding-off of lithium dendrite can be prevented via methods as follows: (a) to design more flexible SEI on the lithium anode surface, which keeps intact during lithium dissolution and deposition; (b) to design current collector with a high specific surface area, decreasing the real current density on the lithium anode surface to confine the formation of lithium dendrite.

5. Develop Li^+ selective conducting membrane to separate the PS based on the principles as follows: (a) size sieving effect; (b) charge repulsion effect; (c) selective adsorption effect, and etc.

6. Develop novel electrolyte, which is stable with anode and could control the diffusion of PS from cathode to anode.

7. Emphasize on the development of Li–S primary batteries. Lithium sulfur battery has already achieved a specific energy over 900 Wh/kg, which is promising for practical applications.

3   Research and Development of Li–O_2 Batteries

3.1   General Introduction

The research of Li–O_2 battery originated in the 1970s, when the oil crisis broke out. At that time, the electrolyte is based on water, which corrodes the lithium heavily. Until 1996, Abraham and Jiang used the gel electrolyte to improve the lithium stability [135], which is considered as the establishment of electrochemical Li–O_2 system. However, the Li–O_2 battery did not attract enough attention at that time, only J. Read in the army lab continued the related research. In 2006, P.G. Bruce group improved the charge/discharge performance of Li-air battery [136], with a much higher specific energy than the commercial Li-ion battery, which attracted global interests.

   In 2012, the International Business Machines Corporation (IBM) launched the "Battery 500" program to develop Li–O_2 battery for electric vehicles which could run 500 miles on a single charge. It's deemed as the landmark project of Li–O_2 research in the 21st century. In 2013, the Toyota Motor Corp. in Japan and the BMW AG in Germany signed a contract to co-develop lithium-air battery, which is deemed as the next-generation battery technology. New Energy and Industrial Technology Development Organization (NEDO) in Japan and the Department of Energy (DOE) in the USA also provided a lot of funding to promote the research of the Li-Air battery. At present, the major institutions around the world involved in the Li–O_2 battery research include the University of St Andrews, University of Waterloo, the Pacific Northwest National Laboratory, Massachusetts Institute of Technology, Argonne National Laboratory, Chinese Academy of Sciences, and etc.

Based on the statistics of the Web of Knowledge, there exist more than 6600 research papers on lithium battery, among them over 1800 papers are from labs in China.

However, the Li–O₂ battery is still premature for commercial application. And there are many important scientific and technical issues yet to be resolved. All these issues will be discussed here.

3.2   Principle and Clarification of Li–O₂ Batteries

The basic principle of Li–O₂ battery is shown in Fig. 22. Li–O₂ battery uses metal lithium as the anode active material, oxygen in the air as the cathode active material, and conductive agent containing a soluble lithium salt as the electrolyte. During the discharge process, the lithium metal loses electrons and becomes lithium ions, which is similar to the Li–S batteries. The lithium ions and electrons conduct through the electrolyte and external circuit respectively to the air cathode, forming Li_2O_2 [137, 138] or LiOH with oxygen. The electricity power is exported via external circuit during the discharge process. During charge, the process is vice versa.

Similar to the classification of Li–S batteries, the Li–O₂ battery systems could also be classified into 4 types: the aprotic organic electrolyte (Fig. 23a), all-solid electrolyte (Fig. 23b), aqueous electrolyte (Fig. 23c) and organic–aqueous hybrid electrolyte (Fig. 23d). These systems all have very serious problems in addition to common problems such as lithium dendrite, unstable SEI and electrolyte decomposition, thus limiting their industrial development. As for the aprotic organic electrolyte, the discharge product, Li_2O_2, would accumulate at the cathode surface to block the reactive sites and prematurely end the discharge process. Besides that, the electrolyte decomposition also occurs at the cathode, due to the high charge

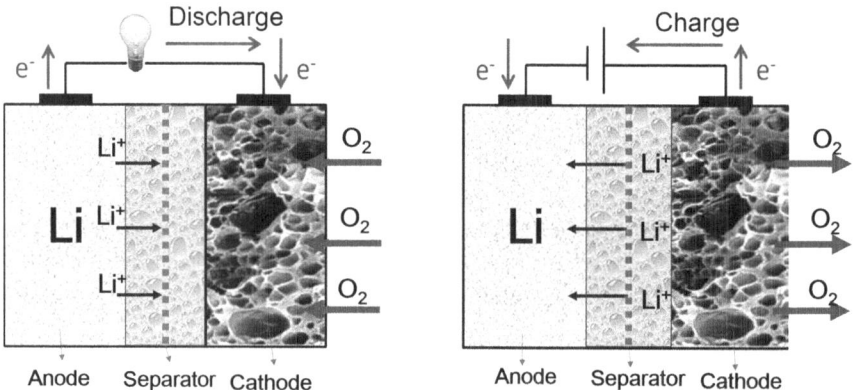

Fig. 22 The work principle of Li–O₂ battery

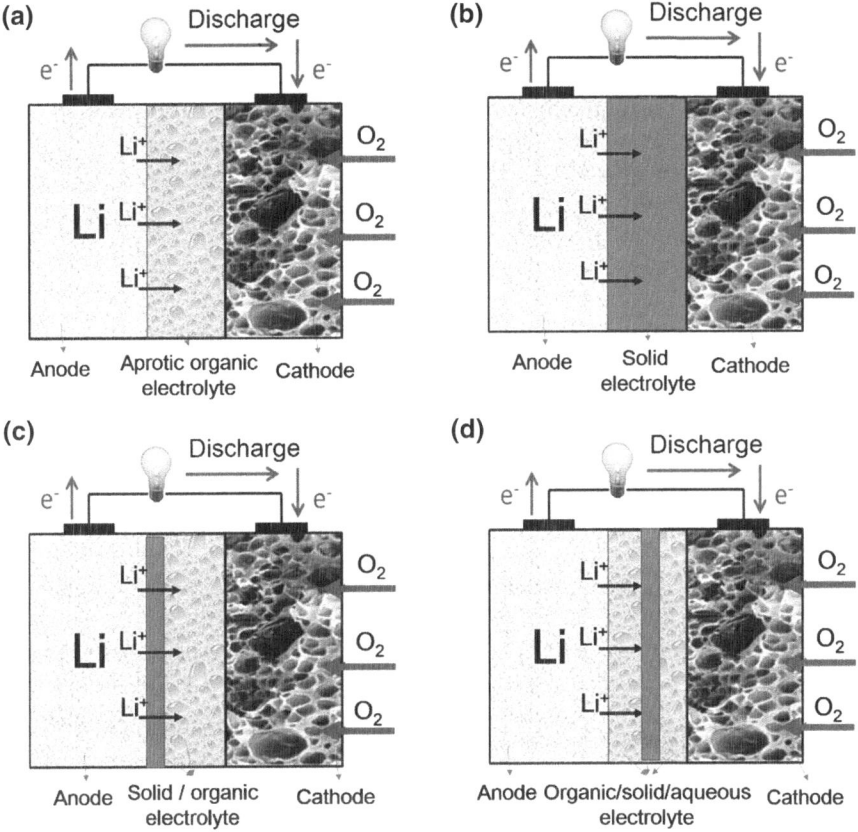

Fig. 23 The schematic illustration of four types of Li–O₂ batteries

voltage (over 4.5 V vs. Li) and the reaction between electrolyte and the oxidative Li₂O₂. As for the organic-aqueous hybrid electrolyte, the discharge product LiOH is dissolved in water, which could avoid the blockage of the reaction sites. This type of battery was proposed by Polyplus Co. in 2007, and was developed by Zhou et al. However, the major problem is its strong dependence on the solid electrolyte separator for perfectly separating water from organic electrolyte, which currently is unable to achieve large-scale production and has low Li⁺ conductivity. All-solid electrolyte battery is aimed to solve the problem of organic electrolyte decomposition, and provide excellent safety and high temperature performance. However, the all-solid electrolyte batteries not only face the problem of cathode blocking by solid product Li₂O₂, but also suffer from the difficulty of developing solid electrolyte with high Li⁺ conductivity. As for the aqueous electrolyte battery, the lithium needs to be tightly protected with a layer of dense solid electrolyte, which leads to great resistance at the interface between lithium and solid electrolyte during charge and discharge.

The above-described four types of Li–O$_2$ battery systems have their own advantages and limitations, depending on the development of key materials and structural design of the battery. Now the Li–O$_2$ battery is still at the early stage of development, lacking thorough understanding of the basic reaction mechanism and some tough technical problems. In recent years, researchers have mainly focused on the development of electrolytes, lithium anodes, cathodes, breathable waterproof and CO$_2$-proof membranes. All these would be discussed as below based on the aprotic organic electrolyte battery.

3.3 Research Status of Li–O$_2$ Batteries

3.3.1 Electrolyte

The electrolytes of Li–O$_2$ battery play important roles in dissolving oxygen and conducting lithium ions, to ensure stable and high energy output. Currently, the electrochemical stability of the as-developed electrolyte is poor on the cathode and anode, which is a major bottleneck that restricts the development of Li–O$_2$ battery.

Before 2010, carbonate electrolyte was mainly used in Li–O$_2$ battery. However, Mizuno [139], Bruce and McCloskey et al. [140, 141] demonstrated that the carbonate electrolyte is vulnerable to the attack of O$_2\bullet$ during operation, causing severe decomposition. Ethers electrolyte shows higher stability to O$_2\bullet$ than the carbonate electrolyte, but the electrolyte decomposition still occurs with byproducts such as organic lithium salt or Li$_2$CO$_3$. When the cycle continues, the cycle stability of lithium-air batteries would be reduced.

In addition to these two kinds of electrolytes, the acetonitrile (ACN) [142–144], dimethyl sulfoxide (DMSO) [145–147], dimethylformamide (DMA) [148], anisole [149] and ionic liquids [150] are also studied. For example, P.G. Bruce's Group [136] reported a very stable lithium-air battery, using 0.1 M LiClO4/DMSO as the electrolyte and porous gold as the cathode. The battery showed a very good cycle performance and capacity retention, with Li$_2$O$_2$ as the main discharge product and trace amount of Li$_2$CO$_3$ and HCO$_2$Li as the byproducts. However, the compatibility between DMSO and lithium anode is not good, which needs further improvement.

Besides the solvents, additives in the Li–O$_2$ battery have also been studied, such as lithium nitrate [151], lithium iodide [152, 153] and tetrathiafulvalene [154]. The main role of these additives is to increase the discharge capacity and decrease the overpotential of battery during charge and discharge. So, the additives could also be seen as the soluble catalysts for cathode reactions. Dr. Bin Xie from the Institute of Physics, Chinese Academic Sciences first discovered that boron-based anion receptor compound, tris (pentafluorophenyl) boron (TPFPB), could promote the dissolution of Li$_2$O$_2$ and Li$_2$O in an organic solvent [155]. Qu et al. [156, 157] found that TPFPB could complex with O$_2\bullet$ to improve the solubility of Li$_2$O$_2$, and reduce the oxidation potential of Li$_2$O$_2$. Bruce et al. [154] found that tetrathiaful-valene (TTF) can be used as a redox mediator, which can significantly reduce the charge overpotential, and improve the rate capability of the battery. At low charge potential, TTF can be oxidized to form TTF$^+$, then the TTF$^+$ could oxide Li$_2$O$_2$ to O$_2$ and become TTF again, facilitating the charging process. Kang et al. [152] found that LiI is a better redox media, which can reduce the charge overpotential more significantly. Recently, scientists at Cambridge University [153] used water and LiI as electrolyte additives, and the ultimate discharge product is the LiOH instead of the ordinary Li$_2$O$_2$. LiOH is more stable than Li$_2$O$_2$, which could greatly reduce the side reactions and improve battery performance. Wherein, addition of LiI helps to break down LiOH, as well as to protect the lithium metal, which enables the battery to work normally even at the presence of an appropriate amount of water.

3.3.2 Lithium Metal Anode Materials

The anode materials of Li–O$_2$ battery and Li–S battery are almost the same. They also face the same problems such as lithium dendrite formation, unstable SEI and electrolyte decomposition. As a result, the research on anode material of Li–O$_2$ battery is also similar to Li–S battery.

3.3.3 Oxygen Cathode

Air cathode of lithium-air batteries is similar to that of other metal-air cells, which consists of a current collector layer, an oxygen diffusion layer and a catalyst layer. The oxygen diffusion layer is mainly used to provide access for O$_2$ to get into the catalyst layer, with a porous layer of hydrophobic polytetrafluoroethylene (PTFE) material to stop the infiltration of electrolyte. Due to the excellent conductivity, oxygen adsorption and reduction activity of carbon, the catalyst layer is mainly composed of carbon materials loaded with catalysts. The catalyst layer not only provides the reaction sites for O$_2$ and Li, but also provides space for Li$_2$O$_2$ deposition and growth. When the carbon surface and pores are totally blocked by Li$_2$O$_2$, the discharge process stops. Therefore, a suitable selection of carbon material and optimization of the physical structure is a key factor to improve the performance of lithium-air battery.

Carbon Materials

Currently, the porous carbon material is the most widely studied air cathode material. This is not only because it has high electrical conductivity to provide fast charge transfer, but also because it has large surface area and low density to improve the specific energy of Li–O$_2$ battery. Besides that, the defect sites in porous carbon could also play a catalytic role during charge and discharge process. The most commonly used porous carbon material in O$_2$ cathode is activated carbon, including Super P [158–161], Ketjen Black [162], Vulcan XC-72 [143], BP2000 and so on. Besides that, a lot of novel materials such as carbon nano-fiber [163], carbon nano-tube [164, 165] and graphene [166], have also been used as the cathode materials of Li–O$_2$ battery. For example, Scrosati et al. [158] coated Super P and binder onto carbon paper as the catalytic layer, and achieved a discharge capacity of 5000 mAh/g under the current density of 3 A/g (based on carbon mass), which was tested with R2032 coin cell and LiCF$_3$SO$_3$/TEGDME electrolyte. Xiao et al. [167] used the hierarchical porous graphene as the catalyst layer, delivering 15,000 mAh/g (C) capacity without any other catalyst. This excellent performance is due to the unique surface structure and property of the hierarchical porous graphene: large pores to facilitate O$_2$ diffusion, nano pores to provide a large quantity of reactive sites and surface defects to accelerate the nucleation.

However, although the carbon material shows excellent capacity delivery during discharge, it could react with Li$_2$O$_2$ and form Li$_2$CO$_3$, which is difficult to be utilized during charge. Non-carbon materials, such as Au, TiC, Co$_x$O$_y$ and Ni, own relatively good electrochemical stability, but they would decrease the practical specific energy. For example, Xinbo Zhang's group [168] constructed a binder-free 3D porous graphene electrode on porous nickel, delivering the capacity of 11,060 and 2020 mAh/g under the current density of 280 and 2800 mA/g (C).

Catalyst

Without a catalyst, the energy efficiency of Li–O$_2$ battery is usually less than 70 %, which is much lower than that of the lithium-ion battery. This is due to the relatively large potential difference between charging and discharging, which is usually about 2.7 V during discharge and 4.2 V during charge. High charge voltage can also cause decomposition of the electrolyte and the electrode materials. Therefore, the development of high-performance, high stability, low-cost cathode ORR/OER catalyst for improving the energy efficiency and cycle life of the Li–O$_2$ battery is critical.

Up to now, some precious metals, transition metal oxides and transition metal complexes are all found with excellent catalytic activity in Li–O$_2$ battery. Precious metals including platinum and platinum-based materials have been widely used as the catalyst for various kinds of reactions, especially as the oxygen reduction catalyst for fuel cell, with high catalytic activity and selectivity, and great resistance against oxidation and corrosion. Since the lithium-air battery cathode is similar to the regenerative fuel cell cathode, precious metals also show good performance in Li–O$_2$ battery. For example, Shaohorn et al. [169] found that 40 wt% Au on XC72 carbon (Au/C) could catalyze the ORR and OER of Li–O$_2$ battery, with the discharge voltage plateau of 2.7 V and the charge voltage plateau of 3.8 V. Furthermore, the Au–Pt catalyst could decrease the OER voltage to 3.4–3.8 V. Besides that, Ru and RuO$_2$ also show good catalytic efficiency when loaded on the nano carbon, carbon nanotube, porous graphene and LTO with different morphologies. According to Guoxiu Wang's work [170], the potential difference of charge and discharge could be decreased to 0.355 V and below.

Although the precious metals own excellent catalytic performance in ORR and OER, they are too expensive for large-scale application. Due to the relatively low price and good catalytic activity, the transition metal oxides are expected to become a substitute for precious metal catalysts. In 2007, Bruce et al. [171] found that Fe$_3$O$_4$, CuO and CoFe$_2$O$_4$ show good catalytic performance. In particular, Co$_3$O$_4$ was found to improve the discharge specific capacity as well as cycle stability of the battery [172]. Although the electrolyte decomposition cannot be avoided, these research results are still very meaningful.

Besides the catalysts mentioned above, the research of other metal composites as the Li–O$_2$ battery catalysts has also made significant progress, such as metal complexes [173–178], metal nitrides [179–182], spinel-type [183–188], pyrochlore-type [189] and perovskite-type [190–194] metal oxides, and mixed metal sulfides [195, 196]. Abraham et al. [66] reported the first cobalt phthalo-cyanine on carbon as an O$_2$ cathode catalyst, which exhibits similar performance to the precious metal catalyst, with the discharge voltage plateau at 2.8 V and charge voltage plateau at 3.7 V. Shui et al. [177] developed the Fe/N/C composites with uniform atomic distribution as the lithium-air battery catalyst. It was found that only O$_2$ gas was evolved during the charge process, while CO$_2$ evolution occurred when using MnO$_2$/XC-72 as a catalyst layer. It means that there is little electrolyte decomposition when Fe/N/C composite catalyst is used.

Until now, a variety of different catalysts have been developed for Li–O$_2$ battery, however, the working mechanism of the catalyst is still unclear. An efficient bifunctional catalyst should not only be able increase the battery capacity, but also be able to reduce the over potential and improve the energy efficiency of each cycle. Besides that, the design and optimization of electrode structure at the micro- and nanoscale is still urgently needed, in order to provide more space for the deposition of discharge products, as well as to ensure the effective transfer of the reactants and the stable electrochemical reaction interface.

Breathable Waterproof and CO$_2$-Proof Membrane

The Li–O$_2$ battery is deemed to get the oxygen gas from the real air environment. As a result, the harmful gases, such as H$_2$O and CO$_2$, must be excluded from air. As for the aprotic electrolyte Li–O$_2$ system, both H$_2$O and CO$_2$ must be excluded from air, because they would react with lithium and Li$_2$O$_2$ to form harmful byproducts. Even for aqueous electrolyte Li–O$_2$ system, the CO$_2$ could still react with the LiOH to form Li$_2$CO$_3$, and lead to the battery failure. Zhang's group [162] used a polymer film to seal the lithium-air batteries, which could deliver a specific energy of 362 Wh/kg (based on the total mass of the battery) during a month, in the environment of 0.21 bar partial pressure of oxygen and 20 % relative humidity. Crowther et al. [197] used a Teflon-coated glass fiber membrane (TCFC) as oxygen selective membrane, which could provide a high enough oxygen permeation rate to achieve the 0.2 mA/cm^2 current density during discharge. In addition, after more than 40 days, only 2 % liquid electrolyte was volatized, and the overpotential of lithium metal anode increased by only 13–24 mV. The anode appears bright like a new one, which showed that the moisture was effectively prevented from entering the battery. Aishui Yu's group synthesized a breathable waterproof membrane using proton-doped polyaniline (PAN), which could obtain the discharge capacity of 3241 mAh/g, at 20 % relative humidity and 0.1 mA/cm^2 current density. Haoshen Zhou's group [198] used less volatile and hydrophobic ionic liquid to replace the traditional organic electrolyte, by adding a hydrophobic diffusion layer between the catalyst layer and the carbon paper collector, they achieved a capacity up to 10,730 mAh/g when operated in air, which provided important ideas for the transformation from Li–O$_2$ battery to Li-air battery.

3.4   Further Research Directions of Li–O$_2$ Battery

Due to its super high specific energy, the secondary lithium-air battery has attracted great attention in recent years. Since the electrolyte changed from the carbonate to the ether system, the detection and characterization method of Li$_2$O$_2$ has been rapidly developed. Now, Li$_2$O$_2$ as a major discharge product has been confirmed in several electrolyte systems (such as TEGDME, DMSO, PP13TFSI ionic liquids, and etc.), and a preliminary understanding of the mechanism of nucleation, growth and decomposition also has been obtained. Based on these studies, the cycle life of a secondary lithium-air battery can be increased to tens or even hundreds of times, with well-controlled discharge capacity. However, as an efficient long-life secondary battery, the Li–O$_2$ battery is still far from practical application. There are still several scientific issues to be addressed [7]:

1. To deepen the understanding of the growth and decomposition mechanism of Li$_2$O$_2$ or LiOH under different voltages, current densities, catalysts and electrolytes.
2. To clarify the formation and decomposition mechanism of by-product Li$_2$CO$_3$ in the aprotic organic electrolyte Li–O$_2$ system, such as its coexistent structure with Li$_2$O$_2$, spatial distribution, growth direction, which could build up a foundation for quantitatively detecting and eliminating the accumulation of Li$_2$CO$_3$.
3. To understand the formation and suppression mechanism of lithium dendrites, which can relief the safety risks caused by battery short circuit.
4. To understand the decomposition mechanism of lithium in organic electrolyte at both anode and cathode.
5. To comprehensively understand the reason of over potential under different stages of charge and discharge, such as the ohmic polarization, mass transfer polarization and electrochemical polarization. It needs a quantitative study based

on the whole electrode, especially with high area-loading of Li_2O_2, in order to achieve a high practical specific energy.

6. To comprehensively understand the catalytic mechanism of ORR and OER in organic electrolyte Li–O₂ system, which might be different from that in the aqueous environment. Besides, the catalysts that could accelerate the decomposition of Li_2CO_3 are also of great interest.
7. To design the Li–O₂ batteries with high specific energy. Most reported batteries are based on the cumbersome laboratory apparatus, which could not exhibit the high specific energy of Li–O₂ batteries.

Solving these issues will play a crucial role in substantially increasing the performance of Li–O₂ battery. It is believed that through persistent efforts and collaborative research, the applications of secondary Li–O₂ batteries will be promoted continually. Then, the secondary Li–O₂ batteries with high specific energy and long life could be finally realized.

3.4.1 Recommended Research Directions

Overall, Li–O₂ batteries are still at the early stage of development, and they are still far away from commercialization. IBM research team pointed out that "as a car power battery, it takes about 35 years of research and development from the nickel metal hydride batteries to lithium-ion battery. It should also have a similar research and development course for lithium—air batteries from research to application." If we can effectively solve these existing problems and promote the research of Li–O₂ batteries into practical applications, it will effectively promote the development of traffic, consumer electronics, aerospace, military, renewable energy and other areas, thereby bringing immeasurable economic benefits. The recent research should be focused on the directions discussed in following subsections.

Cathode Materials with High Stability and Catalytic Activity

In the cathode reaction of an organic lithium-air battery system, Li_2O_2 first deposits and then decomposes, with discharge capacity, energy density and cycle life heavily depending on the structure and surface properties of the cathodes. The following work should be done to improve the cathode stability:

1. Cover the unstable sites of carbon materials with stable compounds Al_2O_3 or Ti_3N_4, which could be used to resist the corrosion of Li_2O_2.
2. Improve the stability of the interface by hetero-atom-modification, for example, by replacing the oxygen atom with nitrogen, sulfur, boron or other atoms to modify the physicochemical properties of carbon surface.
3. Develop non-carbon electrode materials with high specific surface area and large pore volume.

As for the catalyst materials, Ru and RuO_2 show much better performance than others, however, the loading amount of Ru based catalyst is too large (over 20 % in mass). Therefore below work needs to be done:

1. Decrease the loading of precious catalysts;
2. Develop non-precious catalyst.

Electrolyte Materials with High Stability and Catalytic Activity

The voltage window of Li–O₂ battery is almost over 4.5 V, rendering most electrolyte materials such as esters, ethers or sulfone, unstable on anode or cathode. The strategies to develop highly stable electrolyte are as follows:

1. To clarify the mechanism of electrolyte decomposition caused by Li_2O_2, $O_2\cdot$ and high charge potential, and to develop stable electrolyte such as ionic liquid;
2. Develop soluble redox medium more effective than LiI, which could further decrease the over potential during charge and decrease the electrochemical decomposition on cathode;
3. Develop effective additives to promote the formation of stable SEI on lithium anode surface;
4. Develop all-solid electrolyte or gel electrolyte with high stability, high Li^+ conductivity and high oxygen solubility, to overcome the difficulty of electrolyte development.

Anode Materials with High Stability

The batteries using lithium anode always face the same problem, which is the formation and shedding off of lithium dendrite. The major research topics are as follows:

1. Form flexible SEI on lithium surface, which is not easy to break down during charge and discharge;
2. Design anode current collectors with high specific surface area to decrease the current density of lithium deposition;
3. Develop anode materials which could be inserted with Li^+ and with high performance.

High-performance Separator

The solid electrolyte membrane can effectively block the transport of water molecules from the cathode to the anode, which is demanded by all four types of Li–O₂ battery systems. However, the currently developed solid electrolyte membranes always have the problems of high cost, crisp, low conductivity and short cycle life, which are of great importance to be resolved.

Selective Permeation of Oxygen from Air

Li–O_2 battery is a semi-open system. Moisture, carbon dioxide and other harmful gases in the air might enter into the battery and lead to the battery failure. So, selective permeation of oxygen is a key issue for the practical use of Li–O_2 battery. In order to achieve this goal, research should be focused on:

1. Seal the battery in a membrane with high O_2 selectivity and flux, which could stop the H_2O and CO_2 at the same time;
2. Develop the aqueous Li–O_2 battery or hybrid Li–O_2 battery;
3. Develop a pretreatment technology to remove harmful gases in air via chemical or physical methods.

4 Applications and Demonstrations

4.1 The Demonstrations of Li–S Batteries

4.1.1 The Demonstration of Li–S Batteries in Unmanned Aerial Vehicles (UAV)

As an important technology, the commercial demonstration of Li–S batteries is inevitable. The most successful demonstration of Li–S batteries was operated by the Sion Power Company, which is considered as a global leader in this field. In 2009, the Sion Power Company and the Airbus Defence and Space Company successfully flew a High Altitude Pseudo-Satellite, setting world records for endurance and altitude. The vehicle flew for two weeks at an average altitude of 70,000 ft. without landing or refueling. By day, it flies on solar power and recharges batteries. By night, it is powered by Sion Power's Li–S battery. The parameters of the battery stack are compared with Li-ion battery in Table 2.

From this demonstration, it is clarified that the Li–S batteries have already met the technical requirements of the UAVs, compared to Li-ion batteries. However, although 2 weeks is already the longest endurance time for the unrefueled UAVs in the world, it could be further prolonged to months once the cycle life of the Li–S batteries is further increased. In other words, the flight endurance of the UAVs is mainly limited by the cycle life of Li–S batteries.

Table 2 The comparison of parameters of Li–S battery developed by Sion Power Company and Li-ion battery

Total 25.2 V 6.6 Ah (167 Wh) pack		
	Li-Ion battery	Li–S battery
Configuration	7S3P	12S3P
Cell capacity	2.2 Ah	2.2 Ah
Pack weight (g)	1075	640
Wh/kg (pack level)	155	260

4.1.2 The Demonstration of Li–S Batteries in Other Fields

Compared to unrefueled UAVs, the demonstration of Li–S batteries in other potential areas, such as electric vehicles or consumer electronics, still lacks inspiring reports. Even through, great efforts have already been engaged in these areas (Fig. 24).

In 2016, Dalian Institute of Chemical Physics has demonstrated the application of a 12 KWh Li–S battery stack to solar energy storage, supported by the Ministry of Science and Technology of People's Republic of China. However, the cycle life of this kind of battery is still below 100 times, which still cannot meet the requirement of practical application.

As for electric vehicle application, the Li–S battery has the potential to be an enabling technology for both Plug In Hybrid and Pure Electric Vehicles. The specific energy of over 350 Wh/Kg today exceeds most manufacturers' expectations for specific energy of a vehicle battery. Li–S battery improves not only the driving range of the vehicle, but also the overall payload capability. With a compromise between the battery volume and battery weight, Li–S vehicle battery pack can achieve one of the primary goals set forth by automobile manufacturers—a driving range of 300 miles. Therefore, it is easy to understand why Li–S battery is the future power source for electric vehicles. Until now, however, there is still no demonstration of Li–S batteries in this area, due to the limitation of cycle life.

Even though, the Li–S battery still owns great potential for military application, which has much lower-level requirement in cycle life. In addition, the primary Li–S battery with super high specific energy is urgently needed for the military portable power. In 2016, Zhang's group at Dalian Institute of Chemical Physics has already manufactured Li–S batteries with a specific energy over 900 Wh/kg and 1000 Wh/L (Fig. 25).

4.2 The Demonstration of Li–O$_2$ Batteries

The typical demonstration of Li–O$_2$ batteries is carried out by Polyplus Battery Company, who achieved 800 Wh/kg in air and 1300 Wh/kg in water. However, the

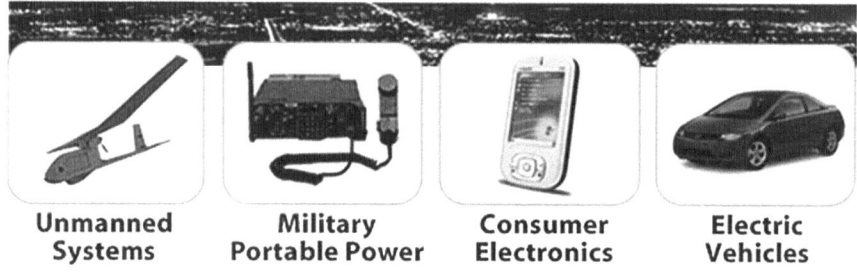

Unmanned Systems   **Military Portable Power**   **Consumer Electronics**   **Electric Vehicles**

Fig. 24 The possible applications of Li–S batteries

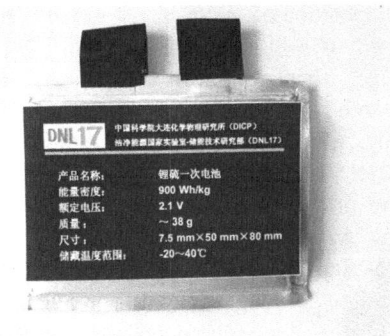
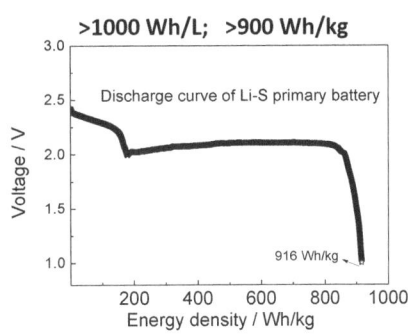

Fig. 25 The Li–S primary battery developed at Dalian Institute of Chemical Physics

demonstrated Li–O$_2$ battery could not be charged yet. Even though, the super high specific energy of the Li–O$_2$ battery already renders it possible to be used in water or other specific areas.

Xinbo Zhang's group at Changchun Institute of Applied Chemistry has also demonstrated a 5 Ah primary Li–O$_2$ battery, which achieved 526 Wh/kg under 0.05 °C at room temperature. The Li–O$_2$ battery with a capacity of 51 Ah and a specific energy of 360 Wh/kg was also developed.

Up to now, the Li–O$_2$ battery technology still cannot meet the technical requirements of electric vehicles, which is even more premature compared to Li–S batteries. However, the Li–O$_2$ battery owns much higher specific energy than Li–S battery, especially for use in aqueous environment where the discharge products would dissolve into water. In general, there is still a long way to go to realize the large-scale application of Li–O$_2$ batteries.

5 Challenges and Perspectives

To put it in a nutshell, although the Li–S and Li–O$_2$ batteries have been studied for decades, there are still many unsolved scientific and technical problems. And these remaining problems are all difficult to overcome by the currently available technologies. Even though, the Li–S and Li–O$_2$ batteries are still important technologies to be developed in the future, due to their super high specific energy compared to Li-ion batteries.

Primary batteries of Li–S and Li–O$_2$ should be first developed to make good use of the advantage of high specific energy, as well as to avoid the disadvantage of short cycle life. Then, these batteries could first find their potential applications in both the military and the civilian fields. However, in order to promote the practical application of secondary Li–S and Li–O$_2$ batteries, both the application demonstration and fundamental research should be put forward at the same time. There are several common problems or challenges for Li–S and Li–O$_2$ batteries, such as the safety issue of lithium anode and the decomposition of electrolyte. As for the Li–O$_2$ battery, how to remove H$_2$O and CO$_2$ from air is also a tough problem.

However, these problems will be solved finally in the future, based on the continuous progress achieved in this research area, especially the development of anodes, electrodes, separators and cathodes. Currently, many companies such as Sion Power Co. (USA), Oxis Energy Co. (British) and Polyplus (USA) claim that they have achieved breakthroughs in the protection of lithium anode, and the performance of Li–S and Li–O$_2$ batteries would be increased gradually. Many powerful Li-ion battery companies, such as Samsung and Sony, also invested into the development of Li–S batteries, which might accelerate the commercialization of Li–S and Li–O$_2$ batteries. However, there is still a long journey to go before the Li–S and Li–O$_2$ batteries could replace the commercial Li-ion batteries in electric vehicles.

References

1. Turner JA (1999) A realizable renewable energy future. Science 285(5428):687–689. doi:10.1126/science.285.5428.687
2. Bruce PG, Freunberger SA, Hardwick LJ, Tarascon JM (2012) Li–O$_2$ and Li–S batteries with high energy storage. Nat Mater 11(1):19–29. doi:10.1038/nmat3191
3. Dunn B, Kamath H, Tarascon JM (2011) Electrical energy storage for the grid: a battery of choices. Science 334(6058):928–935. doi:10.1126/science.1212741
4. Xiao J (2015) Understanding the lithium sulfur battery system at relevant scales. Adv Energy Mater 5(16). doi:10.1002/aenm.201501102
5. Herbert D UJ (1962) Electric dry cells and storage battery US Pat, 3043896
6. Kim HS, Jeong C-S, Kim Y-T (2011) Shuttle inhibitor effect of lithium perchlorate as an electrolyte salt for lithium–sulfur batteries. J Appl Electrochem 42(2):75–79. doi:10.1007/s10800-011-0373-1
7. Bauer I, Thieme S, Brückner J, Althues H, Kaskel S (2014) Reduced polysulfide shuttle in lithium–sulfur batteries using Nafion-based separators. J Power Sources 251:417–422. doi:10.1016/j.jpowsour.2013.11.090
8. Diao Y, Xie K, Xiong S, Hong X (2013) Shuttle phenomenon—the irreversible oxidation mechanism of sulfur active material in Li–S battery. J Power Sources 235:181–186. doi:10.1016/j.jpowsour.2013.01.132
9. Wang X, Wang Z, Chen L (2013) Reduced graphene oxide film as a shuttle-inhibiting interlayer in a lithium–sulfur battery. J Power Sources 242:65–69. doi:10.1016/j.jpowsour.2013.05.063
10. Mikhaylik YV, Akridge JR (2004) Polysulfide shuttle study in the Li/S battery system. J Electrochem Soc 151(11):A1969. doi:10.1149/1.1806394
11. Rosenman A, Elazari R, Salitra G, Markevich E, Aurbach D, Garsuch A (2015) Effect of interactions and reduction products of LiNO$_3$, the anti-shuttle agent, in Li–S battery systems. J Electrochem Soc 162(3):A470–A473. doi:10.1149/2.0861503jes
12. Gu M, Lee J, Kim Y, Kim JS, Jang BY, Lee KT, Kim B-S (2014) Inhibiting shuttle effect in lithium-sulfur battery using layer-by-layer assembled ion-permselective separator. RSC Adv. doi:10.1039/c4ra09718a
13. Gu M, Lee J, Kim Y, Kim JS, Jang BY, Lee KT, Kim B-S (2014) Inhibiting the shuttle effect in lithium-sulfur batteries using a layer-by-layer assembled ion-permselective separator. RSC Adv 4(87):46940–46946. doi:10.1039/c4ra09718a

14. Togasaki N, Momma T, Osaka T (2016) Enhanced cycling performance of a Li metal anode in a dimethylsulfoxide-based electrolyte using highly concentrated lithium salt for a lithium–oxygen battery. J Power Sources 307:98–104. doi:10.1016/j.jpowsour.2015.12.123

15. Zhong Y, Yang M, Zhou X, Luo Y, Wei J, Zhou Z (2015) Orderly packed anodes for high-power lithium-ion batteries with super-long cycle life: rational design of $MnCO_3$/large-area graphene composites. Adv Mater 27(5):806–812. doi:10.1002/adma.201404611

16. Yamada T, Ito S, Omoda R, Watanabe T, Aihara Y, Agostini M, Ulissi U, Hassoun J, Scrosati B (2015) All solid-state lithium-sulfur battery using a glass-type P2S5-Li2S electrolyte: benefits on anode kinetics. J Electrochem Soc 162(4):A646–A651. doi:10.1149/2.0441504jes

17. Qian J, Henderson WA, Xu W, Bhattacharya P, Engelhard M, Borodin O, Zhang J-G (2015) High rate and stable cycling of lithium metal anode. Nat Commun 6. doi:10.1038/ncomms7362

18. Li NW, Yin YX, Yang CP, Guo YG (2015) An artificial solid electrolyte interphase layer for stable lithium metal anodes. Adv Mater 28(9):1853–1858. doi:10.1002/adma.201504526

19. Han Y, Duan X, Li Y, Huang L, Zhu D, Chen Y (2015) Effects of sulfur loading on the corrosion behaviors of metal lithium anode in lithium–sulfur batteries. Mater Res Bull 68:160–165. doi:10.1016/j.materresbull.2015.03.042

20. Fan K, Tian Y, Zhang X, Tan J (2015) Application of stabilized lithium metal powder and hard carbon in anode of lithium–sulfur battery. J Electroanal Chem. doi:10.1016/j.jelechem.2015.10.020

21. Cheng X-B, Zhang Q (2015) Dendrite-free lithium metal anodes: stable solid electrolyte interphases for high-efficiency batteries. J Mater Chem A 3(14):7207–7209. doi:10.1039/c5ta00689a

22. Cao R, Xu W, Lv D, Xiao J, Zhang J-G (2015) Anodes for rechargeable lithium-sulfur batteries. Adv Energy Mater n/a-n/a. doi:10.1002/aenm.201402273

23. Camacho-Forero LE, Smith TW, Bertolini S, Balbuena PB (2015) Reactivity at the lithium-metal anode surface of lithium-sulfur batteries. J Phys Chem C. doi:10.1021/acs.jpcc.5b08254

24. Zhou X (2014) Sustained room-temperature sodium-ion battery anodes using graphene-templated carbon hybrid. J Phys Chem

25. Zhang X, Wang W, Wang A, Huang Y, Yuan K, Yu Z, Qiu J, Yang Y (2014) Improved cycle stability and high security of Li-B alloy anode for lithium–sulfur battery. J Mater Chem A 2(30):11660. doi:10.1039/c4ta01709a

26. Xiong S, Xie K, Diao Y, Hong X (2014) Characterization of the solid electrolyte interphase on lithium anode for preventing the shuttle mechanism in lithium–sulfur batteries. J Power Sources 246:840–845. doi:10.1016/j.jpowsour.2013.08.041

27. Ma G, Wen Z, Wu M, Shen C, Wang Q, Jin J, Wu X (2014) A lithium anode protection guided highly-stable lithium-sulfur battery. Chem Commun 50(91):14209–14212. doi:10.1039/C4CC05535G

28. Kummer JT, Weber N (1968) Sodium-sulfur secondary battery. Sae Trans 76:88-&

29. Fally J, Lasne C, Lazennec Y, Lecars Y, Margotin P (1972) Study of a beta-alumina electrolyte for sodium sulfur battery. J Electrochem Soc 119(3):C110-&

30. Fally J, Lasne C, Lazennec Y, Margotin P (1972) Some aspects of sodium-sulfur battery working. J Electrochem Soc 119(3):C110-&

31. Silverma HP, Seo ET, Gelb GH, Richards NA (1972) Load-leveling applications of sodium-sulfur batteries in large power-plants. J Electrochem Soc 119(8):C214

32. Tischer RP (1972) Sodium-sulfur battery. J Electrochem Soc 119(3):C110-&

33. Fally J, Lasne C, Lazennec Y, Lecars Y, Margotin P (1973) Study of a beta-alumina electrolyte for sodium-sulfur battery. J Electrochem Soc 120(10):1296–1298

34. Whalen TJ, Tennenho GJ, Meyer C (1973) Influence of composition and microstructure on properties of beta-alumina conductive ceramics for sodium-sulfur battery. Am Ceram Soc Bull 52(4):435–436

35. Jones IW (1974) Sodium-sulfur batteries for traction in United Kingdom. Abstracts of Papers of the American Chemical Society, pp 89–89
36. Weiner SA, Janz GJ, Gordon RS (1974) Sodium-sulfur battery—government, industry, university project. Am Ceram Soc Bull 53(8):602
37. Whalen TJ, Meyer C (1974) Properties and microstructure of conductive ceramics for sodium-sulfur battery. Am Ceram Soc Bull 53(4):344
38. http://www.sionpower.com/index.php (2016) Accessed 3 Aug 2016
39. http://www.polyplus.com/ (2016) Accessed 3 Aug 2016
40. http://www.oxisenergy.com/ (2016) Accessed 3 Aug 2016
41. http://english.dicp.cas.cn/ns_17179/ue/201511/t20151126_156499.html (2016) Accessed 3 Aug 2016
42. Zhang S (2012) Improved cyclability of liquid electrolyte lithium/sulfur batteries by optimizing electrolyte/sulfur ratio. Energies 5(12):5190–5197. doi:10.3390/en5125190
43. Aurbach D, Granot E (1997) The study of electrolyte solutions based on solvents from the "glyme" family (linear polyethers) for secondary Li battery systems. Electrochim Acta 42 (4):697–718. doi:10.1016/s0013-4686(96)00231-9
44. Granot D (1996) The study of electrolyte solutions based on solvents from the ccglyme" family (linear polyethers) for secondary Li battery systems
45. Ryu H-S, Ahn H-J, Kim K-W, Ahn J-H, Cho K-K, Nam T-H, Kim J-U, Cho G-B (2006) Discharge behavior of lithium/sulfur cell with TEGDME based electrolyte at low temperature. J Power Sources 163(1):201–206. doi:10.1016/j.jpowsour.2005.12.061
46. Ryu HS, Ahn HJ, Kim KW, Ahn JH, Cho KK, Nam TH (2006) Self-discharge characteristics of lithium/sulfur batteries using TEGDME liquid electrolyte. Electrochim Acta 52(4):1563–1566. doi:10.1016/j.electacta.2006.01.086
47. Zhang S, Ueno K, Dokko K, Watanabe M (2015) Recent advances in electrolytes for lithium-sulfur batteries. Adv Energy Mater n/a-n/a. doi:10.1002/aenm.201500117
48. Lin Z (2015) Developments of electrolyte systems for lithium-sulfur batteries: a review. Front Energy Res. doi:10.3389/fenrg.2015.00005
49. Jo G, Jeon H, Park MJ (2015) Synthesis of polymer electrolytes based on poly(ethylene oxide) and an anion-stabilizing hard polymer for enhancing conductivity and cation transport. ACS Macro Lett 4(2):225–230. doi:10.1021/mz500717j
50. Zhaoyin W (2014) A shuttle effect free lithium sulfur battery based on a hybrid electrolyte. doi:10.1039/C4CP03694H, 10.1039/c0xx00000x, 10.1039/b000000x
51. Scheers J, Fantini S, Johansson P (2014) A review of electrolytes for lithium–sulphur batteries. J Power Sources 255:204–218. doi:10.1016/j.jpowsour.2014.01.023
52. Zhang SS (2013) New insight into liquid electrolyte of rechargeable lithium/sulfur battery. Electrochim Acta 97:226–230. doi:10.1016/j.electacta.2013.02.122
53. Zhang SS (2013) Liquid electrolyte lithium/sulfur battery: fundamental chemistry, problems, and solutions. J Power Sources 231:153–162. doi:10.1016/j.jpowsour.2012.12.102
54. Chen R, Liu Z, Li L, Wu F (2013) Electrolyte materials for high energy density lithium-sulfur secondary battery. Chin Sci Bull (Chinese Version) 58(32):3301. doi:10.1360/972013-661
55. Zhao Y, Zhang Y, Gosselink D, Doan TNL, Sadhu M, Cheang H-J, Chen P (2012) Polymer electrolytes for lithium/sulfur batteries. Membranes 2(4):553–564. doi:10.3390/membranes2030553
56. Teran AA, Balsara NP (2011) Effect of lithium polysulfides on the morphology of block copolymer electrolytes. Macromolecules 44(23):9267–9275. doi:10.1021/ma202091z
57. Peled E, Sternberg Y, Gorenshtein A, Lavi Y (1989) Lithium-Sulfur battery: evaluation of dioxolane-based electrolytes. J Electrochem Soc 136(6):1621–1625
58. Chang DR, Lee SH, Kim SW, Kim HT (2002) Binary electrolyte based on tetra(ethylene glycol) dimethyl ether and 1,3-dioxolane for lithium–sulfur battery. J Power Sources 112 (2):452–460
59. Mikhaylik Y, Tucson A (2014) Electrolytes for lithium sulfur cells. US Patent 8,828,610 B2

60. Xiong S, Kai X, Hong X, Diao Y (2011) Effect of LiBOB as additive on electrochemical properties of lithium–sulfur batteries. Ionics 18(3):249–254. doi:10.1007/s11581-011-0628-1

61. Lin Z, Liu Z, Fu W, Dudney NJ, Liang C (2013) Phosphorous pentasulfide as a novel additive for high-performance lithium-sulfur batteries. Adv Funct Mater 23(8):1064–1069. doi:10.1002/adfm.201200696

62. Suo L, Hu Y-S, Li H, Armand M, Chen L (2013) A new class of solvent-in-salt electrolyte for high-energy rechargeable metallic lithium batteries. Nat Commun 4. doi:10.1038/ncomms2513

63. Yuan LX, Feng JK, Ai XP, Cao YL, Chen SL, Yang HX (2006) Improved dischargeability and reversibility of sulfur cathode in a novel ionic liquid electrolyte. Electrochem Commun 8 (4):610–614. doi:10.1016/j.elecom.2006.02.007

64. Hassoun J, Scrosati B (2010) Moving to a solid-state configuration: a valid approach to making lithium-sulfur batteries viable for practical applications. Adv Mater 22(45):5198–5201. doi:10.1002/adma.201002584

65. Rajendran S (2004) Li-ion conduction of plasticized PVA solid polymer electrolytes complexed with various lithium salts. Solid State Ionics 167(3–4):335–339. doi:10.1016/j.ssi.2004.01.020

66. Kim S, Jung Y, Lim HS (2004) The effect of solvent component on the discharge performance of lithium–sulfur cell containing various organic electrolytes. Electrochim Acta 50(2–3):889–892. doi:10.1016/j.electacta.2004.01.093

67. Belostotskii AM, Markevich E, Aurbach D (2004) On Li-chelating additives to electrolytes for Li batterres. J Coord Chem 57(12):1047–1056. doi:10.1080/00958970412331281809

68. Aurbach D, Talyosef Y, Markovsky B, Markevich E, Zinigrad E, Asraf L, Gnanaraj JS, Kim HJ (2004) Design of electrolyte solutions for Li and Li-ion batteries: a review. Electrochim Acta 50(2–3):247–254. doi:10.1016/j.electacta.2004.01.090

69. Aurbach D, Schechter A (2004) Advanced liquid electrolyte solutions. Lithium batteries: science and technology, pp 530–573

70. Kamaya N, Homma K, Yamakawa Y, Hirayama M, Kanno R, Yonemura M, Kamiyama T, Kato Y, Hama S, Kawamoto K, Mitsui A (2011) A lithium superionic conductor. Nat Mater 10(9):682–686. doi:http://www.nature.com/nmat/journal/v10/n9/abs/nmat3066.html#supplementary-information

71. Capiglia C, Imanishi N, Takeda Y, Henderson WA, Passerini S (2003) Poly(ethylene oxide) LiN(SO[sub 2]CF[sub 2]CF[sub 3])[sub 2] polymer electrolytes. J Electrochem Soc 150(4): A525. doi:10.1149/1.1557963

72. Wang JL, Yang J, Xie JY, Xu NX, Li Y (2002) Sulfur–carbon nano-composite as cathode for rechargeable lithium battery based on gel electrolyte. Electrochem Commun 4(6):499–502. doi:10.1016/S1388-2481(02)00358-2

73. Shin JH, Jung SS, Kim KW, Ahn HJ, Ahn JH (2002) Preparation and characterization of plasticized polymer electrolytes based on the PVdF-HFP copolymer for lithium/sulfur battery. J Mater Sci: Mater Electron 13(12):727–733. doi:10.1023/a:1021521207247

74. Unemoto A, Yasaku S, Nogami G, Tazawa M, Taniguchi M, Matsuo M, Ikeshoji T, S-i Orimo (2014) Development of bulk-type all-solid-state lithium-sulfur battery using LiBH4 electrolyte. Appl Phys Lett 105(8):083901. doi:10.1063/1.4893666

75. Nagao M, Imade Y, Narisawa H, Kobayashi T, Watanabe R, Yokoi T, Tatsumi T, Kanno R (2013) All-solid-state Li–sulfur batteries with mesoporous electrode and thio-LISICON solid electrolyte. J Power Sources 222:237–242. doi:10.1016/j.jpowsour.2012.08.041

76. Marmorstein D (2002) Solid state lithium/sulfur batteries for electric vehicles: electrochemical and spectroelectrochemical investigations

77. Zhang Y, Zhao Y, Gosselink D, Chen P Synthesis of poly(ethylene-oxide)/nanoclay solid polymer electrolyte for all solid state lithium/sulfur battery. Ionics

78. marmorstein d solid state lithium/sulfur batteries for electric vehicles: electrochemical and spectroelectrochemical investigations

79. Kobayashi T, Imade Y, Shishihara D, Homma K, Nagao M, Watanabe R, Yokoi T, Yamada A, Kanno R, Tatsumi T (2008) All solid-state battery with sulfur electrode and thio-LISICON electrolyte. J Power Sources 182(2):621–625. doi:10.1016/j.jpowsour.2008. 03.030

80. Song JH, Yeon JT, Jang JY, Han JG, Lee SM, Choi NS (2013) Effect of fluoroethylene carbonate on electrochemical performances of lithium electrodes and lithium-sulfur batteries. J Electrochem Soc 160(6):A873–A881. doi:10.1149/2.101306jes

81. Langenhuizen NPW (1998) The effect of mass transport on Li deposition and dissolution. J Electrochem Soc 145(9):3094–3099

82. Gan H, Takeuchi ES (1996) Lithium electrodes with and without CO$_2$ treatment: electrochemical behavior and effect on high rate lithium battery performance. J Power Sources 62:45–50

83. Ishikawa M, Yoshitake S, Morita M, Matsuda Y (1994) In situ scanning vibrating electrode technique for the characterization of interface between lithium electrode and electrolytes containing additives. J Electrochem Soc 141(12):L159–L161

84. Ding F, Xu W, Graff GL, Zhang J, Sushko ML, Chen X, Shao Y, Engelhard MH, Nie Z, Xiao J, Liu X, Sushko PV, Liu J, Zhang JG (2013) Dendrite-free lithium deposition via self-healing electrostatic shield mechanism. J Am Chem Soc 135(11):4450–4456. doi:10. 1021/ja312241y

85. Agostini M, Scrosati B, Hassoun J (2015) An advanced lithium-ion sulfur battery for high energy storage. Adv Energy Mater n/a-n/a. doi:10.1002/aenm.201500481

86. Zhang SS (2012) Role of LiNO$_3$ in rechargeable lithium/sulfur battery. Electrochim Acta 70:344–348. doi:10.1016/j.electacta.2012.03.081

87. Jozwiuk A, Berkes BB, Wei Sommer H, Janek J, Brezesinski T (2016) The critical role of lithium nitrate in the gas evolution of lithium-sulfur batteries. Energy Environ Sci. doi:10. 1039/C6EE00789A

88. Aurbach D, Pollak E, Elazari R, Salitra G, Kelley CS, Affinito J (2009) On the surface chemical aspects of very high energy density, rechargeable Li–Sulfur batteries. J Electrochem Soc 156(8):A694. doi:10.1149/1.3148721

89. Ji X, Lee KT, Nazar LF (2009) A highly ordered nanostructured carbon-sulphur cathode for lithium-sulphur batteries. Nat Mater 8(6):500–506. doi:10.1038/nmat2460

90. Xin S, Gu L, Zhao NH, Yin YX, Zhou LJ, Guo YG, Wan LJ (2012) Smaller sulfur molecules promise better lithium-sulfur batteries. J Am Chem Soc 134(45):18510–18513. doi:10.1021/ja308170k

91. Yan Y, Yin YX, Xin S, Guo YG, Wan LJ (2012) Ionothermal synthesis of sulfur-doped porous carbons hybridized with graphene as superior anode materials for lithium-ion batteries. Chem Commun (Camb) 48(86):10663–10665. doi:10.1039/c2cc36234a

92. Yin YX, Xin S, Guo YG, Wan LJ (2013) Lithium-sulfur batteries: electrochemistry, materials, and prospects. Angew Chem Int Ed Engl 52(50):13186–13200. doi:10.1002/anie. 201304762

93. Pang Q, Tang J, Huang H, Liang X, Hart C, Tam KC, Nazar LF (2015) A nitrogen and sulfur dual-doped carbon derived from Polyrhodanine@Cellulose for advanced lithium-sulfur batteries. Adv Mater 27(39):6021–6028. doi:10.1002/adma.201502467

94. Tang C, Zhang Q, Zhao MQ, Huang JQ, Cheng XB, Tian GL, Peng HJ, Wei F (2014) Nitrogen-doped aligned carbon nanotube/graphene sandwiches: facile catalytic growth on bifunctional natural catalysts and their applications as scaffolds for high-rate lithium-sulfur batteries. Adv Mater 26(35):6100–6105. doi:10.1002/adma.201401243

95. Wu F, Li J, Tian Y, Su Y, Wang J, Yang W, Li N, Chen S, Bao L (2015) 3D coral-like nitrogen-sulfur co-doped carbon-sulfur composite for high performance lithium-sulfur batteries. Sci Rep 5:13340. doi:10.1038/srep13340

96. Liu J, Li W, Duan L, Li X, Ji L, Geng Z, Huang K, Lu L, Zhou L, Liu Z, Chen W, Liu L, Feng S, Zhang Y (2015) A graphene-like oxygenated carbon nitride material for improved cycle-life lithium/sulfur batteries. Nano Lett 15(8):5137–5142. doi:10.1021/acs.nanolett. 5b01919

97. Chung WJ, Griebel JJ, Kim ET, Yoon H, Simmonds AG, Ji HJ, Dirlam PT, Glass RS, Wie JJ, Nguyen NA, Guralnick BW, Park J, Somogyi A, Theato P, Mackay ME, Sung Y-E, Char K, Pyun J (2013) The use of elemental sulfur as an alternative feedstock for polymeric materials. Nat Chem 5(6):518–524. doi:10.1038/nchem.1624

98. Kim H, Lee J, Ahn H, Kim O, Park MJ (2015) Synthesis of three-dimensionally interconnected sulfur-rich polymers for cathode materials of high-rate lithium-sulfur batteries. Nat Commun 6:7278. doi:10.1038/ncomms8278

99. Pang Q, Nazar LF (2016) Long-life and high-areal-capacity Li–S batteries enabled by a light-weight polar host with intrinsic polysulfide adsorption. ACS Nano 10(4):4111–4118. doi:10.1021/acsnano.5b07347

100. Wei Seh Z, Li W, Cha JJ, Zheng G, Yang Y, McDowell MT, Hsu P-C, Cui Y (2013) Sulphur–TiO$_2$ yolk–shell nanoarchitecture with internal void space for long-cycle lithium–sulphur batteries. Nat Commun 4:1331. doi:10.1038/ncomms2327

101. Zhou W, Yu Y, Chen H, DiSalvo FJ, Abruña HD (2013) Yolk-Shell structure of polyaniline-coated sulfur for lithium-sulfur batteries. J Am Chem Soc 135(44):16736–16743. doi:10.1021/ja409508q

102. Zhang SS (2012) Binder based on polyelectrolyte for high capacity density lithium/sulfur battery. J Electrochem Soc 159(8):A1226–A1229. doi:10.1149/2.039208jes

103. Jin Z, Xie K, Hong X (2013) Electrochemical performance of lithium/sulfur batteries using perfluorinated ionomer electrolyte with lithium sulfonyl dicyanomethide functional groups as functional separator. RSC Adv 3(23):8889. doi:10.1039/c3ra41517a

104. Zhang Z, Lai Y, Zhang Z, Zhang K, Li J (2014) Al$_2$O$_3$-coated porous separator for enhanced electrochemical performance of lithium sulfur batteries. Electrochim Acta 129:55–61. doi:10.1016/j.electacta.2014.02.077

105. Yao H, Yan K, Li W, Zheng G, Kong D, Seh ZW, Narasimhan VK, Liang Z, Cui Y (2014) Improved lithium-sulfur batteries with a conductive coating on the separator to prevent the accumulation of inactive S-related species at the cathode-separator interface. Energy Environ Sci 7(10):3381–3390. doi:10.1039/c4ee01377h

106. Wei H, Ma J, Li B, Zuo Y, Xia D (2014) Enhanced cycle performance of lithium-sulfur batteries using a separator modified with a PVDF-C layer. ACS Appl Mater Interfaces 6 (22):20276–20281. doi:10.1021/am505807k

107. Chung S-H, Manthiram A (2014) Bifunctional separator with a light-weight carbon-coating for dynamically and statically stable lithium-sulfur batteries. Adv Funct Mater 24(33):5299–5306. doi:10.1002/adfm.201400845

108. Vizintin A, Patel MUM, Genorio B, Dominko R (2014) Effective separation of lithium anode and sulfur cathode in lithium-sulfur batteries. Chemelectrochem 1(6):1040–1045. doi:10.1002/celc.201402039

109. Chung S-H, Manthiram A (2014) High-performance Li–S batteries with an ultra-lightweight MWCNT-coated separator. J Phys Chem Lett 5(11):1978–1983. doi:10.1021/jz5006913

110. Li W, Hicks-Garner J, Wang J, Liu J, Gross AF, Sherman E, Graetz J, Vajo JJ, Liu P (2014) V$_2$O$_5$ polysulfide anion barrier for long-lived Li–S batteries. Chem Mater 26(11):3403–3410. doi:10.1021/cm500575q

111. Li L, Chen Y, Guo X, Zhong B (2015) Preparation of sodium trimetaphosphate and its application as an additive agent in a novel polyvinylidene fluoride based gel polymer electrolyte in lithium sulfur batteries. Polym Chem. doi:10.1039/c4py01353k

112. Wang X, Wang Z, Chen L (2013) Reduced graphene oxide film as a shuttle-inhibiting interlayer in a lithium-sulfur battery. J Power Sources 242:65–69. doi:10.1016/j.jpowsour.2013.05.063

113. Zu C, Su Y-S, Fu Y, Manthiram A (2013) Improved lithium-sulfur cells with a treated carbon paper interlayer. Phys Chem Chem Phys 15(7):2291–2297. doi:10.1039/c2cp43394j

114. Zhang K, Qin F, Fang J, Li Q, Jia M, Lai Y, Zhang Z, Li J (2014) Nickel foam as interlayer to improve the performance of lithium-sulfur battery. J Solid State Electrochem 18(4):1025–1029. doi:10.1007/s10008-013-2351-5

115. Zhang K, Li Q, Zhang L, Fang J, Li J, Qin F, Zhang Z, Lai Y (2014) From filter paper to carbon paper and toward Li–S battery interlayer. Mater Lett 121:198–201. doi:10.1016/j. matlet.2014.01.151

116. Su Y-S, Manthiram A (2012) Lithium-sulphur batteries with a microporous carbon paper as a bifunctional interlayer. Nat Commun 3. doi:1166/ncomms2163

117. Chung S-H, Manthiram A (2014) Carbonized eggshell membrane as a natural polysulfide reservoir for highly reversible Li–S batteries. Adv Mater 26(9):1360–1365. doi:10.1002/adma.201304365

118. Singhal R, Chung S-H, Manthiram A, Kalra V (2015) Free-standing carbon nanofiber interlayer for high-performance lithium-sulfur batteries. J Mater Chem A. doi:10.1039/C4TA06511E

119. Zhang SS (2013) A concept for making poly(ethylene oxide) based composite gel polymer electrolyte lithium/sulfur battery. J Electrochem Soc 160(9):A1421–A1424. doi:10.1149/2.058309jes

120. Zhang Y, Zhao Y, Bakenov Z, Gosselink D, Chen P (2014) Poly(vinylidene fluoride-co-hexafluoropropylene)/poly(methylmethacrylate)/nanoclay composite gel polymer electrolyte for lithium/sulfur batteries. J Solid State Electrochem 18(4):1111–1116. doi:10.1007/s10008-013-2366-y

121. Li M, Yang B, Zhang Z, Wang L, Zhang Y (2013) Polymer gel electrolytes containing sulfur-based ionic liquids in lithium battery applications at room temperature. J Appl Electrochem 43(5):515–521. doi:10.1007/s10800-013-0535-4

122. Jin ZQ, Xie K, Hong XB, Hu ZQ, Liu X (2012) Application of lithiated Nafion ionomer film as functional separator for lithium sulfur cells. J Power Sources 218:163–167. doi:10.1016/j. jpowsour.2012.06.100

123. Jin Z, Xie K, Hong X, Hu Z, Liu X (2012) Application of lithiated Nafion ionomer film as functional separator for lithium sulfur cells. J Power Sources 218:163–167. doi:10.1016/j. jpowsour.2012.06.100

124. Jin Z, Xie K, Hong X (2013) Synthesis and electrochemical properties of a perfluorinated ionomer with lithium sulfonyl dicyanomethide functional groups. J Mater Chem A 1(2):342. doi:10.1039/c2ta00134a

125. Tang Q, Shan Z, Wang L, Qin X, Zhu K, Tian J, Liu X (2014) Nafion coated sulfur–carbon electrode for high performance lithium–sulfur batteries. J Power Sources 246:253–259. doi:10.1016/j.jpowsour.2013.07.076

126. Huang J-Q, Zhang Q, Peng H-J, Liu X-Y, Qian W-Z, Wei F (2014) Ionic shield for polysulfides towards highly-stable lithium–sulfur batteries. Energy Environ Sci. doi:10.1039/c3ee42223b

127. Ong SP, Chevrier VL, Hautier G, Jain A, Moore C, Kim S, Ma X, Ceder G (2011) Voltage, stability and diffusion barrier differences between sodium-ion and lithium-ion intercalation materials. Energy Environ Sci 4(9):3680. doi:10.1039/c1ee01782a

128. Wang L, Zhang T, Yang S, Cheng F, Liang J, Chen J (2013) A quantum-chemical study on the discharge reaction mechanism of lithium-sulfur batteries. J Energy Chem 22(1):72–77. doi:10.1016/s2095-4956(13)60009-1

129. Atomistic Simulation Group in the Materials Department of Imperial College, Database of Ionic Radii, http://abulafia.mt.ic.ac.uk/shannon/ptable.php (2014) Accessed 4 Dec 2014

130. Yan N, Yang X, Zhou W, Zhang H, Li X, Zhang H (2015) Fabrication of a nano-Li$^+$-channel interlayer for high performance Li–S battery application. RSC Adv 5(33):26273. doi:10.1039/c5ra01269d

131. Wang Q, Yan N, Wang M, Qu C, Yang X, Zhang H, Li X, Zhang H (2015) Layer-by-layer assembled C/S cathode with trace binder for Li–S battery application. ACS Appl Mater Interfaces 7(45):25002–25006. doi:10.1021/acsami.5b08887

132. Aravindan V, Lee Y-S, Madhavi S (2015) Research progress on negative electrodes for practical Li-Ion batteries: beyond carbonaceous anodes. Adv Energy Mater 5(13):1402225. doi:10.1002/aenm.201402225

133. Yuvaraj S, Selvan RK, Lee YS (2016) An overview of AB2O4- and A2BO4-structured negative electrodes for advanced Li-ion batteries. RSC Adv 6(26):21448–21474. doi:10.1039/c5ra23503k

134. Goriparti S, Miele E, De Angelis F, Di Fabrizio E, Proietti Zaccaria R, Capiglia C (2014) Review on recent progress of nanostructured anode materials for Li-ion batteries. J Power Sources 257:421–443. doi:10.1016/j.jpowsour.2013.11.103

135. Abraham KM, Jiang Z (1995) A polymer electrolyte-based rechargeable lithium/oxygen battery. J Electrochem Soc 27(1):1–5

136. Peng Z, Freunberger SA, Chen Y, Bruce PG (2012) A reversible and higher-rate Li–O$_2$ battery. Science 337 (6094):563–566

137. Read J, Read J (2005) Ether-Based electrolytes for the lithium/oxygen organic electrolyte battery. J Electrochem Soc 153(1):A96–A100

138. Takeshi O, Aurelie D, Michael H, Petr N, Bruce PG (2006) Rechargeable Li$_2$O$_2$ electrode for lithium batteries. J Am Chem Soc 128(4):1390–1393

139. Mizuno F, Nakanishi S, Kotani Y, Yokoishi S, Iba H (2010) Rechargeable Li-Air batteries with carbonate-based liquid electrolytes. Electrochemistry 78(5):403–405

140. Freunberger SA, Chen Y, Peng Z, Griffin JM, Hardwick LJ, Bardé F, Novák P, Bruce PG (2011) Reactions in the rechargeable lithium–O$_2$ battery with alkyl carbonate electrolytes. J Am Chem Soc 133(20):8040–8047. doi:10.1021/ja2021747

141. Freunberger SA, Chen Y, Drewett NE, Hardwick LJ, Bardé F, Bruce PG (2011) The lithium-oxygen battery with ether-based electrolytes †. Angew Chem Int Ed 50(37):8609–8613

142. Peng Z, Freunberger SA, Hardwick LJ, Chen Y, Vincent G, Fanny B, Petr N, Duncan G, Jean-Marie T, Bruce PG (2011) Oxygen reactions in a non-aqueous Li$^+$ electrolyte. Angew Chem Int Ed 50(28):6351–6355

143. Mccloskey BD, Bethune DS, Shelby RM, Mori T, Scheffler R, Speidel A, Sherwood M, Luntz AC (2012) Limitations in rechargeability of Li–O$_2$ batteries and possible origins. J Phys Chem Lett 3(20):3043–3047

144. Laoire CO, Mukerjee S, Abraham KM, Plichta EJ, Hendrickson MA Influence of nonaqueous solvents on the electrochemistry of oxygen in the rechargeable lithium—air battery. J Phys Chem C (ACS Publications). American Chemical Society

145. Lopez N, Graham DJ, Jr MGR, Alliger GE, Shaohorn Y, Cummins CC, Nocera DG (2012) Reversible reduction of oxygen to peroxide facilitated by molecular recognition. Science 335 (6067):450–453

146. Sun B, Zhang J, Munroe P, Ahn HJ, Wang G (2013) Hierarchical NiCO$_2$O$_4$ nanorods as an efficient cathode catalyst for rechargeable non-aqueous Li–O$_2$ batteries. Electrochem Commun 31(6):88–91

147. Trahan MJ, Mukerjee S, Plichta EJ, Hendrickson MA, Abraham KM (2013) Studies of Li-Air cells utilizing dimethyl sulfoxide-based electrolyte. J Electrochem Soc 160(2):A259–A267

148. Chen Y, Freunberger SA, Peng Z, Bardé F, Bruce PG (2012) Li–O$_2$ battery with a dimethylformamide electrolyte. J Am Chem Soc 134(18):7952–7957. doi:10.1021/ja302178w

149. Walker W, Giordani V, Uddin J, Bryantsev VS, Chase GV, Addison D (2013) A rechargeable Li–O$_2$ battery using a lithium nitrate/N, N-dimethylacetamide electrolyte. J Am Chem Soc 135(6):2076–2079

150. Allen CJ, Mukerjee S, Plichta EJ, Hendrickson MA, Abraham KM (2011) Oxygen electrode rechargeability in an ionic liquid for the Li–air battery. J Phys Chem Lett 2(19):2420–2424

151. Sharon D, Hirsberg D, Afri M, Chesneau F, Lavi R, Frimer AA, Sun YK, Aurbach D (2015) Catalytic behavior of lithium nitrate in Li–O$_2$ Cells. Acs Appl Mater Interfaces 7(30)

152. Lim HD, Song H, Kim J, Gwon H, Bae Y, Park KY, Hong J, Kim H, Kim T, Kim YH (2014) Superior rechargeability and efficiency of lithium–oxygen batteries: hierarchical air electrode architecture combined with a soluble catalyst. Angew Chem 126(15):4007–4012

153. Liu T, Leskes M, Yu W, Moore AJ, Zhou L, Bayley PM, Kim G, Grey CP (2015) Cycling Li-O$_2$ batteries via LiOH formation and decomposition. Science 350(6260):530–533

154. Chen Y, Freunberger SA, Peng Z, Fontaine O, Bruce PG (2013) Charging a Li–O$_2$ battery using a redox mediator. Nat Chem 5(6):489–494

155. Xie B, Lee HS, Li H, Yang XQ, McBreen J, Chen LQ (2008) New electrolytes using Li$_2$O or Li$_2$O$_2$ oxides and tris(pentafluorophenyl) borane as boron based anion receptor for lithium batteries. Electrochem Commun 10(8):1195–1197. doi:10.1016/j.elecom.2008.05.043

156. Zheng D, Lee HS, Yang XQ, Qu D (2013) Electrochemical oxidation of solid Li$_2$O$_2$ in non-aqueous electrolyte using peroxide complexing additives for lithium–air batteries. Electrochem Commun 28(28):17–19

157. Wu X, Jie X, Wang D, Jian Z, Zhang JG (2010) Effects of nonaqueous electrolytes on the performance of lithium/air batteries. J Electrochem Soc 157(2):A219–A224

158. Jung HG, Hassoun J, Park JB, Sun YK, Scrosati B (2012) An improved high-performance lithium-air battery. Nat Chem 4(7):579–585

159. Lei Y, Lu J, Luo X, Wu T, Du P, Zhang X, Ren Y, Wen J, Miller DJ, Miller JT (2013) Synthesis of porous carbon supported palladium nanoparticle catalysts by atomic layer deposition: application for rechargeable lithium-O$_2$ battery. Nano Lett 13(9):4182–4189

160. Jung H-G, Kim H-S, Park J-B, Oh I-H, Hassoun J, Yoon CS, Scrosati B, Sun Y-K (2012) A transmission electron microscopy study of the electrochemical process of lithium-oxygen cells. Nano Lett 12(8):4333–4335. doi:10.1021/nl302066d

161. Qin Y, Lu J, Du P, Chen Z, Ren Y, Wu T, Miller JT, Wen J, Miller DJ, Zhang Z (2012) In-situ fabrication of porous-carbon-supported [small alpha]-MnO$_2$ nanorods at room temperature: application for rechargeable Li–O$_2$ battery

162. Zhang JG, Wang D, Wu X, Xiao J, Williford RE (2010) Ambient operation of Li/Air batteries. J Power Sources 195(13):4332–4337

163. Mitchell RR, Gallant BM, Thompson CV, Yang SH (2011) All-carbon-nanofiber electrodes for high-energy rechargeable Li–O$_2$ batteries. Energy Environ Sci 4(8):2952–2958

164. Mitchell RR, Gallant BM, Shao-Horn Y, Thompson CV (2013) Mechanisms of morphological evolution of Li$_2$O$_2$ particles during electrochemical growth. J Phys Chem Lett 4 (7):1060–1064. doi:10.1021/jz4003586

165. Gallant BM, Mitchell RR, Kwabi DG, Zhou J, Zuin L, Thompson CV, Yang SH (2012) Chemical and morphological changes of Li–O$_2$ battery electrodes upon cycling. J Phys Chem C 116(39):20800–20805

166. Yoo E, Zhou H (2011) Li–air rechargeable battery based on metal-free graphene nanosheet catalysts. ACS Nano 5(4):3020–3026. doi:10.1021/nn200084u

167. Xiao J, Mei D, Li X, Xu W, Wang D, Graff GL, Bennett WD, Nie Z, Saraf LV, Aksay IA, Liu J, Zhang J-G (2011) Hierarchically porous graphene as a lithium-air battery electrode. Nano Lett 11(11):5071–5078. doi:10.1021/nl203332e

168. Wang ZL, Xu D, Xu JJ, Zhang LL, Zhang XB (2012) Lithium ion batteries: graphene oxide gel-derived, free-standing, hierarchically porous carbon for high-capacity and high-rate rechargeable Li–O$_2$ batteries (Adv. Funct. Mater. 17/2012). Adv Funct Mater 22(17):3699–3705

169. Harding JR, Lu YC, Tsukada Y, Shaohorn Y (2012) Evidence of catalyzed oxidation of Li$_2$O$_2$ for rechargeable Li-air battery applications. Phys Chem Chem Phys 14(30):10540–10546

170. Sun B, Munroe P, Wang G (2013) Ruthenium nanocrystals as cathode catalysts for lithium-oxygen batteries with a superior performance. Sci Rep 3:2247. doi:10.1038/srep02247, http://www.nature.com/articles/srep02247#supplementary-information

171. Débart A, Bao J, Armstrong G, Bruce PG (2007) An O$_2$ cathode for rechargeable lithium batteries: the effect of a catalyst. J Power Sources 174(2):1177–1182. doi:10.1016/j.jpowsour.2007.06.180

172. Lu Y-C, Xu Z, Gasteiger HA, Chen S, Hamad-Schifferli K, Shao-Horn Y (2010) Platinum–gold nanoparticles: a highly active bifunctional electrocatalyst for rechargeable lithium−air batteries. J Am Chem Soc 132(35):12170–12171. doi:10.1021/ja1036572

173. Morozan A, Jousselme B, Palacin S (2011) Low-platinum and platinum-free catalysts for the oxygen reduction reaction at fuel cell cathodes. Energy Environ Sci 4(4):1238–1254

174. Chen Z, Higgins D, Yu A, Zhang L, Zhang J (2011) A review on non-precious metal electrocatalysts for PEM fuel cells. Energy Environ Sci 4(9):3167–3192

175. Jaouen F, Proietti E, Lefèvre M, Chenitz R, Dodelet JP, Wu G, Chung HT, Johnston CM, Zelenay P (2010) Recent advances in non-precious metal catalysis for oxygen-reduction reaction in polymer electrolyte fuel cells. Energy Environ Sci 4(1):114–130

176. Abraham KM, Jiang ZA (1995) A polymer electrolyte-based rechargeable lithium/oxygen battery. J Electrochem Soc 143(1):1–5

177. Shui JL, Karan NK, Balasubramanian M, Li SY, Liu DJ (2012) Fe/N/C composite in Li–O$_2$ battery: studies of catalytic structure and activity toward oxygen evolution reaction. J Am Chem Soc 134(40):16654–16661

178. Zhu C, Choi JY, Wang H, Hui L, Chen Z (2011) Highly durable and active non-precious air cathode catalyst for zinc air battery. J Power Sources 196(7):3673–3677

179. Qi J, Jiang L, Jiang Q, Wang S, Sun G (2010) Theoretical and experimental studies on the relationship between the structures of molybdenum nitrides and their catalytic activities toward the oxygen reduction reaction. J Phys Chem C 114(42):18159–18166

180. Zhang K, Zhang L, Chen X, He X, Wang X, Dong S, Han P, Zhang C, Wang S, Gu L (2013) Mesoporous cobalt molybdenum nitride: a highly active bifunctional electrocatalyst and its application in lithium–O$_2$ batteries. J Phys Chem C 117(2):858–865

181. Li F, Ohnishi R, Yamada Y, Kubota J, Domen K, Yamada A, Zhou H (2013) Carbon supported TiN nanoparticles: an efficient bifunctional catalyst for non-aqueous Li–O$_2$ batteries. Chem Commun 49(12):1175–1177. doi:10.1039/C2CC37042E

182. Chen J, Takanabe K, Ohnishi R, Lu D, Okada S, Hatasawa H, Morioka H, Antonietti M, Kubota J, Domen K (2010) Nano-sized TiN on carbon black as an efficient electrocatalyst for the oxygen reduction reaction prepared using an mpg-C3N4 template. Chem Commun 46 (40):7492–7494

183. Ríos E, Reyes H, Ortiz J, Gautier JL (2005) Double channel electrode flow cell application to the study of HO$_2^-$ production on Mn$_x$Co$_{3-x}$O$_4$ ($0 \le x \le 1$) spinel films. Electrochim Acta 50(13):2705–2711

184. Koninck MD, Poirier SC, Marsan B (2007) CuxCo$_{3-x}$O$_4$ used as bifunctional electrocatalyst II. Electrochemical characterization for the oxygen reduction reaction. J Electrochem Soc 154(4):A381–A388

185. Nikolova V, Iliev P, Petrov K, Vitanov T, Zhecheva E, Stoyanova R, Valov I, Stoychev D (2008) Electrocatalysts for bifunctional oxygen/air electrodes. J Power Sources 185(2): 727–733

186. Du J, Pan Y, Zhang T, Han X, Cheng F, Chen J (2012) Facile solvothermal synthesis of CaMn$_2$O$_4$ nanorods for electrochemical oxygen reduction. J Mater Chem 22(22): 15812–15818

187. Chinnusamy T, Rodionov V, Kühn FE, Reiser O (2011) Rapid room-temperature synthesis of nanocrystalline spinels as oxygen reduction and evolution electrocatalysts. Nat Chem 3(1):79–84

188. Prakash J, Tryk D, Yeager E (1990) Electrocatalysis for oxygen electrodes in fuel cells and water electrolyzers for space applications. J Power Sources 29(90):413–422

189. Akazawa T, Inaguma Y, Katsumata T, Hiraki K, Takahashi T (2004) Flux growth and physical properties of pyrochlore Pb$_2$Ru$_2$O$_{6.5}$ single crystals. J Cryst Growth 271(3–4): 445–449

190. Jin S, Gasteiger H, Yabuuchi N, Goodenough J, Yang SH (2011) Design principles for oxygen reduction activity on perovskite oxides in alkaline environment

191. Yuasa M, Nishida M, Kida T, Yamazoe N, Shimanoe K (2011) Bi-functional oxygen electrodes using LaMnO$_3$/LaNiO$_3$ for rechargeable metal-air batteries. J Electrochem Soc 158(5):A605–A610

192. Yuasa M, Imamura H, Nishida M, Kida T, Shimanoe K (2012) Preparation of nano-LaNiO$_3$ support electrode for rechargeable metal-air batteries. Electrochem Commun 24(10):50–52

193. Takeguchi T, Yamanaka T, Takahashi H, Watanabe H, Kuroki T, Nakanishi H, Orikasa Y, Uchimoto Y, Takano H, Ohguri N (2013) Layered perovskite oxide: a reversible air electrode for oxygen evolution/reduction in rechargeable metal-air batteries. J Am Chem Soc 135(30):11125–11130
194. Ohkuma H, Uechi I, Imanishi N, Hirano A, Takeda Y, Yamamoto O (2013) Carbon electrode with perovskite-oxide catalyst for aqueous electrolyte lithium-air secondary batteries. J Power Sources 223(223):319–324
195. Zhang Z, Wang X, Cui G, Zhang A, Zhou X, Xu H, Gu L (2014) NiCo2S4 sub-micron spheres: an efficient non-precious metal bifunctional electrocatalyst. Nanoscale 6(7):3540–3544
196. Liu Q, Jin J, Zhang J (2013) NiCo2S4@graphene as a bifunctional electrocatalyst for oxygen reduction and evolution reactions. Acs Appl Mater Interfaces 5(11):5002–5008
197. Crowther O, Keeny D, Moureau DM, Meyer B, Salomon M, Hendrickson M (2012) Electrolyte optimization for the primary lithium metal air battery using an oxygen selective membrane. J Power Sources 202(1):347–351
198. Zhang T, Zhou H (2012) From Li–O$_2$ to Li–air batteries: carbon nanotubes/ionic liquid gels with a tricontinuous passage of electrons, ions, and oxygen. Angew Chem Int Ed 51(44):11062–11067. doi:10.1002/anie.201204983